NATURAL LANGUAGE PROCESSING IN HEALTHCARE

As Natural (NLP) gets more and more traction in healthcare applications, there is a growing demand for developing solutions that can understand, analyze, and generate languages that humans can understand. This book showcases the current advances and scenarios of NLP-based solutions for healthcare and low-resource languages.

Natural Language Processing in Healthcare: A Special Focus on Low Resource Languages covers the theoretical and practical aspects as well as ethical and social implications of NLP in healthcare. It showcases the latest research and developments contributing to the rising awareness and importance of maintaining linguistic diversity. The book goes on to present current advances and scenarios based on solutions in healthcare and low-resource languages and identifies the major challenges and opportunities that will impact NLP in clinical practice and health studies.

This book is self-contained, comprehensive, and will be useful to researchers, academicians, technologists, and students.

Innovations in Big Data and Machine Learning
Series Editors: Rashmi Agrawal and Neha Gupta

This series will include reference books and handbooks that will provide the conceptual and advanced reference materials that cover building and promoting the field of Big Data and Machine Learning which will include theoretical foundations, algorithms and models, evaluation and experiments, applications and systems, case studies, and applied analytics in specific domains or on specific issues.

Artificial Intelligence and Internet of Things
Applications in Smart Healthcare
Edited by Lalit Mohan Goyal, Tanzila Saba, Amjad Rehman, and Souad Larabi

Reinventing Manufacturing and Business Processes through Artificial Intelligence
Edited by Geeta Rana, Alex Khang, Ravindra Sharma, Alok Kumar Goel, and Ashok Kumar Dubey

Convergence of Blockchain, AI, and IoT
Concepts and Challenges
Edited by R. Indrakumari, R. Lakshmana Kumar, Balamurugan Balusamy, and Vijanth Sagayan Asirvadam

Exploratory Data Analytics for Healthcare
Edited by R. Lakshmana Kumar, R. Indrakumari, B. Balamurugan, and Achyut Shankar

Information Security Handbook
Edited by Abhishek Kumar, Anavatti G. Sreenatha, Ashutosh Kumar Dubey, and Pramod Singh Rathore

Natural Language Processing In Healthcare
A Special Focus on Low Resource Languages
Edited by Satya Ranjan Dash, Shantipriya Parida, Esaú Villatoro Tello, Biswaranjan Acharya, and Ondřej Bojar

For more information on this series, please visit: https://www.routledge.com/Innovations-in-Big-Data-and-Machine-Learning/book-series/CRCIBDML

NATURAL LANGUAGE PROCESSING IN HEALTHCARE
A Special Focus on Low Resource Languages

Edited by
Satya Ranjan Dash, Shantipriya Parida,
Esaú Villatoro Tello, Biswaranjan Acharya, and
Ondřej Bojar

CRC Press
Taylor & Francis Group
Boca Raton London New York

CRC Press is an imprint of the
Taylor & Francis Group, an **informa** business

First edition published 2023
by CRC Press
6000 Broken Sound Parkway NW, Suite 300, Boca Raton, FL 33487-2742

and by CRC Press
4 Park Square, Milton Park, Abingdon, Oxon, OX14 4RN

CRC Press is an imprint of Taylor & Francis Group, LLC

© 2023 Taylor & Francis Group, LLC

Library of Congress Cataloguing-in-Publication Data
A catalog record has been requested for this book

ISBN: 978-0-367-68539-3 (hbk)
ISBN: 978-0-367-68540-9 (pbk)
ISBN: 978-1-003-13801-3 (ebk)

DOI: 10.1201/9781003138013

Typeset in Times
by MPS Limited, Dehradun

Contents

Preface

As technology enters more and more areas of our lives, including healthcare situations, there is a growing demand for developing automatic solutions that can understand, analyze, and generate languages that humans speak. The case is more challenging for the low-resource languages. Without a special focus on these languages, the quality of the processing would not be sufficient and the health care solution would not reach the majority of the population who could obtain the benefit. The goal of this book is to collect current advances and scenarios of language processing solutions for healthcare and low-resource languages. Data extracted from the clinical text and clinically relevant texts in languages other than English add another dimension to data aggregation. After years of neglect, low-resource languages (be they minority, regional, endangered, or heritage languages) have made it to the scene of computational linguistics, as increased availability of digital devices, which makes the request for digital usability of low-resource languages stronger. Much clinical information is currently contained in the free text of scientific publications and clinical records. For this reason, Natural Language Processing (NLP) has been increasingly impacting biomedical research. NLP researchers face the need to establish clinical text processing in a language other than English, and clinical informatics researchers and practitioners are looking for resources and NLP tools and techniques for their languages to speed up clinical practice and/or investigation.

The primary element in communication is information exchange. People living in less connected areas are often unable to get the kind of information they need, due to various socioeconomical and technological barriers. As a result, they miss out on crucial knowledge required to improve their well-being. Technology pervades all aspects of society and continues to change the way people access and share information, learn and educate, as well as provide and access services in the healthcare sector and others. Language is the main medium through which such transformational technology can be integrated into the socioeconomic processes of a community. NLP and speech systems, therefore, break down barriers and enable users and whole communities with easy access to information and services.

NLP is an active field of research that aims to teach computers to understand human language. Low-resource natural language processing has recently attracted much attention among NLP researchers due to its need and potential. The low-resource languages are languages that have not enough digital data to train robust NLP technologies, and as a result, few or no automated language processing systems exist for them. The research on these languages and building NLP applications for such languages can reinforce the ties between the world and ensure its diversity. The ability to analyze the clinical text in languages other than English opens access to important medical data concerning the cohorts of patients who are treated in countries where English is not the official language. This edition aims to capture NLP developments in healthcare and their applications across scientific disciplines for low-resource languages. It helps students, researchers, and professionals of the NLP community as well as interdisciplinary researchers involved in the field.

Chapter 1 describes a clinical practice by machine translation on low-resource languages. The chapter starts with the history of translation technologies that have played a vital role in the various crisis and relief scenarios as the Haitian earthquake in 2010 and Translators without Borders (TWB), respectively. The recently dissolved Standby Task Force (SBTF) deployed NLP to tackle misinformation during Coronavirus Pandemic 2019. Chapter 2 presents feature analysis and classification of impaired language caused by brain injury. Language impairment occurs from different illnesses, having a variety of causes. This focuses on the analysis of impaired language caused by a traumatic brain injury (TBI), which can vary from aphasia, apraxia, dysarthria, or other sorts of alterations. Chapter 3 represents a review of NLP for mental disorders. In this chapter, the authors provide an overview of NLP applications and datasets dedicated to address problems related to mental health. The chapter focuses on the different applications proposed, the types of data sources these applications use, and the languages they cover.

Chapter 4 presents an interesting example, healthcare NLP infrastructure for the Greek language. The infrastructure was developed initially for the processing of general language, and extended later on to incorporate biomedical texts as well. The infrastructure comprises: (a) components developed de novo to meet the needs of the domain-specific requirements, such as a biomedical corpus, a generic and application-independent medical ontology, and a multi-word term extraction mechanism, (b) general language processing tools that were enhanced for the processing of the corpus, such as tokenization and sentence splitting tools, and a lexicon-based morphosyntactic tagger. Chapter 5 deals with the recognition of medical domain multiword units (MWU) in texts written in Croatian language. The focus is on the automatic recognition of complex MWUs in low resource settings. Chapter 6 developed HealFavor, a chat-based application for healthcare which is extended with machine translation. This application is inherently designed to personalize the interaction between the user and the system. It allows the user to interact with the system as they would interact with a real-life person, and hence its design must provide real-time feedback and deliver precise decisions. Chapter 7 focuses on the development of a machine translation system for promoting the use of a low-resource language in the clinical domain. In this chapter, the authors describe the approach of developing an MT system for translating clinical text from Basque into Spanish.

Chapter 8 represents the study of various approaches proposed by researchers for detecting and extracting Adverse Drug Reactions (ADRs) from clinical reports, electronic health records, patient narratives, patient's social media queries, and posts. Chapter 9 proposes that methods for detecting fake news so far have assumed that all content in a document is deceptive; however, those texts can include truthful claims. This study shows that the performance of fake news detection can increase if the text is first automatically summarized. The summarization process removes secondary ideas from documents and deceptive claims then stand out clearer for the subsequent classification as truthful or deceptive. The classification results on datasets in Arabic and English languages show an F-measure over 92%. The last chapter presents a setting employing NLP tools to improve patient-provider secure email communication via machine translation.

Editor Biographies

Dr. Satya Ranjan Dash is a computer professional, with his research interest in machine learning, deep learning with NLP, Computational Biology, and Biomedical domain. He is currently working as an associate professor at KIIT University, India. His current research includes Natural Language Processing, particularly text summarization, topic detection, language detection, machine translation for low resource languages.

Dr. Shantipriya Parida is working as a Senior AI Scientist at Silo AI, Finland since Feb 2022. Before Silo, he was working as a post-doctoral researcher at Idiap Research Institute, Switzerland. His current research includes natural language processing, particularly text summarization, topic detection, language detection, machine translation for Indian languages, multi-modal machine translation, and corpus development for low resource Indian languages. Before Idiap, he was working as a Post-doctoral researcher with Prof. Ondřej Bojar at Charles University, Prague working on machine translation, and deep learning. He has expertise in machineLearning, AI, computational neuroscience, product development, system/ solution architecture. He is an organizing committee member for Workshop on Asian Translation WAT2019, WAT2020, WAT2021, and WAT2022. Along with Prof. Ondřej Bojar, he organized the WAT Multimodal-Translation Task for many Indian languages (Hindi, Malayalam, and Bengali). He is part of the program committee/ organizer for many top-tier NLP conferences and workshops.

Dr. Esaú Villatoro Tello holds a tenure position at the Universidad Autónoma Metropolitana campus Cuajimalpa (UAM-C) in Mexico City. Currently, he is an academic visitor at Idiap Research Institute in Martigny Switzerland. His main research interests are related to Natural Language Processing, particularly authorship analysis, and non-thematic text categorization. He is an active member of several research groups and NLP organizations: the Language and Reasoning research Group at UAM-C, the Laboratory of Language Technologies at the National Institute of Astrophysics, Optics and Electronics, the Mexican Association for Natural Language Processing (AMPLN), the Hispano-American Network for Automatic Human Language Processing (RedHisTAL) and the Mexican Academy of Computer Science (AMEXCOMP).

Biswaranjan Acharya is an academic currently associated with Kalinga Institute of Industrial Technology Deemed to be University along with pursuing a Ph.D. in computer application from Veer Surendra Sai University of Technology (VSSUT), Burla, Odisha, India. He has received MCA in 2009 from IGNOU, New Delhi, India and M.Tech in Computer Science and Engineering in the year of 2012 from Biju Pattanaik University of Technology (BPUT), Odisha, India. He is also associated with various educational and research societies like IEEE, IACSIT, CSI, IAENG, and ISC. He has two years of industry experience as a software engineer, a total of ten years of experience in both academia of some reputed universities like Ravenshaw University and the software development field. He is currently working on research area multiprocessor scheduling along with different fields like Data Analytics, Computer Vision, Machine Learning, and IoT. He has more than 50 patents on his credit both national and international. He published some research articles in internationally reputable journals as well as serving as a reviewer.

Dr. Ondřej Bojar is a lead scientist in the field of machine translation in the Czech Republic. He works as an associate professor at the Institute of Formal and Applied Linguistics at Charles University. Machine translation has been in the center of his research interests since 2005, early in his Ph.D. studies. He has been regularly participating and since 2013 co-organizing WMT shared tasks with a specific focus on translation into Czech. His system has dominated English-Czech translation in the years 2013–2015, before deep learning and neural networks fundamentally changed the field. Ondřej's main focus now is a little broader and entails machine learning in general with explicit aims towards meaning representation and natural language understanding, including speech processing.

1 A Clinical Practice by Machine Translation on Low Resource Languages

Rupjyoti Baruah and Anil Kumar Singh
Department of Computer Science and Engineering, Indian
Institute of Technology, Varanasi, India

CONTENTS

1.1 INTRODUCTION: MEDICAL TRANSLATION

Machine Translation (MT) is a sub-field of computational linguistics that automatically translates words or phrases of one human language into another. MT is widely applied to the medical field due to the current growth in interest and success of new language technologies. Medical translation involves the communication of knowledge dealing with various specialties, such as psychology, sociology, pharmacology, psychiatry, and surgery. It is a specialized field of translation in providing healthcare assistance to minorities or foreigners. Hospitals across the country constantly require translation services to ensure fair treatment and correct diagnoses of patients' particular problems. It is crucial to clarify the aspects of an ailment with a doctor and a doctor to fully understand their treatment details. Medical translators are responsible for translating *patient records or medical-legal documents,* hospitals' informational brochures, instructions of use for the medical equipment into a second language. In addition, translators prepare medical files for patients who seek medical advice outside their country of residence. Medical translators require much attention to accurately translate various documents ranging from labels to

DOI: 10.1201/9781003138013-1

TABLE 1.1

Rise of clinical text towards application

Timeline	Digitalization of Clinical Text
800 BCE to ~ 200 CE	Divine and mystical → observation and reasoning
1452–1516	Printed in Greek medical texts
1000–1800 AD	Greek and Arabic terminology by transliteration
End of 1800 AD	Local languages retaining the Græco-Latin
1964–1966	ELISA as a psychotherapist

brochures, medical and patient journals, training materials, and patents. Clinical document translation must be both medically precise and culturally sensitive. Medical professionals rely on medical translations to diagnose patients and monitor the treatment progress of their foreign-speaking patients. Translators have hand-picked clinicians and doctors who understand the content based on the first inter-action with patients locally. Medical translators have related education and work experience in clinical laboratories, hospitals, community health centers, nursing homes, doctors' offices, blood donation centers, and other healthcare facilities. The misuse of medical terminology, inaccurate translation, or carelessness when trans-lating medical reports and medical records can have serious consequences. The keys to producing a successful translation are a "lover of language, attentiveness, a hos-pitality to pursue mysterious terminology, and caring enough to get it exactly right".

Medical jargon is full of sequences of words and idioms, which may sound unusual in everyday speech. Scientific language is a long-standing partner for clinical research organizations. A survey of medical translators (O'Neill 1998) reported that translators are not physicians but specialize in medical translation. Furthermore, the study re-vealed that medical translators acquire background knowledge by joining different medical courses, studying medication, working in a situation directly or indirectly related to healthcare, or participating in medical translation courses. Translation re-quires more than translating phrases in one language for another, adhering to grammar rules, and choosing the appropriate register. A translator has an excellent command of both the source and target languages and needs to be very sensitive to written and implied words. In the literature, several levels are passed to reach the current medical–digital applications. They are increasingly being developed in health pre-vention, diagnostics, and therapy to promise great benefits and potential. Table 1.1 depicts a timeline with the corresponding rise in digitalization of clinical text.

1.2   IMPORTANCE OF MT SYSTEM AND LOW RESOURCE LANGUAGES

MT plays an essential role in a society where different languages are spoken. It removes the language barrier and digital division in society by providing access to all the local languages that a person can understand.

During the Haitian earthquake in 2010, translation technologies played a crucial role in various crisis and relief scenarios that act as disaster prevention and management. Air, land, and sea transport facilities, communication systems, hospitals, and electrical networks were damaged by the earthquake, which hampered early rescue and aid efforts. The earthquake caused an urgent need for outside rescuers to communicate with Haitians whose only language is Haitian Creole. As a result, a mobile translation program to translate between English and Haitian Creole was quickly written (Lewis 2010). Microsoft research developed a web-based English/ Creole translator on the Internet, adding disaster-specific words and phrases to the database. Building a more robust system, Microsoft regularly updates more parallel sentences and phrases in the system by taking medical terminology and other emergency-type notifications and translating them into Haitian-Creole. In addition, Microsoft Translator's extensive API provides support to other software and Web sites. Developers trained an MT engine (Lewis 2010) by searching parallel data compiled by linguists (Rogl 2017), pre-translated medical terms for the rescue teams as Creole to English emergency text messages.

The Covid-19 pandemic showed that the NLP and telehealth technology are not just passing trends in the medical industry. On the contrary, Covid-19 has hit businesses like never before, and the health sector has found it very hard to cope with this sudden change in reality. However, it saw challenges and opportunities which it did not see in the last few decades. As a result, the healthcare industry has transformed rapidly in the last decade. Patient data management is now electronically managed using Electronic Health Records (EHR) or Electronic Medical Records (EMR). With the Covid-19 pandemic and its consequences of lockdowns, the medical industry took responsive measures as lucky enough to benefit from the outbreak, such as supermarkets and the home health with fitness niche. Medical translation has shaped the staggering progress and collective international effort to deal with the coronavirus outbreak made throughout 2020. Coronavirus outbreak has multiplied the logistical barriers for medical interpretation. Medical interpreters must work remotely, multiplying the challenges for front-line doctors and non-English-speaking patients. These issues are not unique to Covid-19. Recently, Ebola outbreaks in Africa and natural disasters in Haiti have seen localized problems. Those worst affected by such issues do not necessarily speak the same language as aid workers and national organizations. Thus, it makes a case for increased attention to language translation in crisis communication.

One of the most pressing challenges has been delivering health advice and guidelines to the people in their native language. However, the issue persists for people residing in different countries when they are not native speakers of the national language. The medical translation practices and translation technology allow us to close these language gaps.

1.2.1 Low Resource Language

Low-resource languages are those that have relatively fewer data available for training conversational AI systems. For example, among 7011 world languages, there are still several languages that are native to a sizable number of people but which may not have considerable amounts of data sets for training an AI model. The importance of providing accurate and human-developed low-resource translations ensures that they do not leave behind individuals with limited English proficiency in response to natural disasters for public safety. MT is currently being developed in the clinical field to improve patient-provider and patient-staff communication in multilingual clinical settings and increase access to health education resources in low-resource languages.

India is a multilingual country as most people speak and understand more than one language or dialect that uses a different script. A famous aphorism depicts India's linguistic diversity: "कोस कोस पर बदले पानी और चार कोस पर वाणी" (Every 3 km (approximately 1 "kos"), the taste of water changes, every 12 km, the language). Articles 344 (O'Neill 1998) and 351 of the Constitution of India, titled the Eighth Schedule, recognizes 22 languages as official languages of the states of India. India is home to more than 19,560 languages or dialects and nearly 97% population in the country call one language included in the 22 scheduled languages as their mother tongue. The remaining 3% speak other languages, according to the Census 2011[1]. With a 121 crore population, 121 languages are spoken by 10,000 or more people in the country.

Language is a critical element of culture and language diversity increases the cultural richness and beauty of linguistic diversity of literature from these different languages. The state of Assam is a gateway of India's North Eastern Region close to its international borders with Bangladesh and Bhutan. Assamese, recognized as an official language of Assam, a branch of Indo-Aryan language, is the easternmost Indo-European language, spoken by over 14 million speakers and serves as the lingua franca of the region. Assamese is an anglicized form of the actual name Asamiya (অসমীয়া). The sister language of Assamese are (Bengali, Maithili, and Oriya) developed from Magadhi Prakrit. The Assamese script has a total of 52 characters with 41 consonants and 11 vowels similar to the Devnagari.

1.2.2   STRUCTURE OF MEDICAL WORD-FORMATION

Most anatomical and clinical terms used in medicine today are Latin or Latinized Greek words, the origin of which can be traced back to the fifth century BC. The physician in ancient Rome or Greece communicated in native languages. Latin was the pre-dominant language used in medicine until the 18th century. Following are a few examples of Greek and Latin prefixes and suffixes with their meaning in Table 1.2 (Fischbach 1998; DžUGANOVá 2013; Karwacka 2015; (42)). When it splits the whole term into its components, then it readily grasps the meaning. As an example, hypoglycemia broken down into hypo (below normal), glyc (sugar), and emia (blood) indicates an insufficient blood sugar level.

Medical terms are similar to learning a new language, like jigsaw puzzles. Terminologies are constructed of small pieces that make each word unique, but the pieces can be used in different combinations in other words as well (Chabner 2020). Studying medical terms analyzes the words by breaking them into parts;

TABLE 1.2

Prefixes and suffixes with their meanings

Prefix	Meaning	Suffix	Meaning
a-	absence of	-algia	pain
brady-	slow	-ectasia	dilatation
dys-	difficult	-ectomy	excision
hyper-	above normal	-emia	blood

terminology is related to the human body's structure and function, and identical pronunciation with a different meaning. The formation of medical word analysis is shown below:

- Morphological through derivation, compounding, abbreviation.
- Medical terms have pretty regular morphology, derived from Greek and Latin languages. The root words are combined with prefixes (start of a word) and suffixes (end of a word). The vowel "o" acts as a connection to the prefix to root words. Medical terms can contain multiple root words in various combinations:
  - Myocardium(মায়'কাৰ্ডয়িাম) = myo- (prefix) + card(ium) (root)
  - Endocarditis(এণ্ড'কাৰ্ডাইটছি) = endo- (prefix) + card (root) + -itis (suffix)
  - Cytology(চাইট'লজী) = cyt(o) (root) + -logy (suffix))
  - Gastroenterology(গেষ্ট্ৰ'এণ্টেৰ'লজী) = gastr(o) (root) + enter(o) (root) + -logy (suffix)
  - Adenoma(এডেন'মা)= aden(o) (root) + oma (suffix)
  - Hydroxynitrodihydrothymine(হাইড্ৰ'ক্সনিাইট্ৰ'ডাইহাইড্ৰ'থায়মাইন) = Hydro (root) + xy (suffix) + nitro (root) + di (suffix) + hydro (root) + thy (suffix) + mine (suffix)
  - Hydroxywybutine (হাইড্ৰ'ক্সবিউটীন) = Hydro (root) + xy (suffix) + wy (suffix) + butine (suffix)
- Most medical terms are compound words made up of prefixes, suffixes and may include multiple roots. Examples are blood donor, blood pressure, and blood group, etc.
- An abbreviation is a contracted form of a word or phrase such as (AIDS (এইড্ছ), HIV (এইচ.আই.ভি.), and Covid (ক'ভিড)).
- Collocation: co-occurrence or combination of words on the syntagmatic level.
  - collocations: haematostasia
  - synonymic variations: myeloproliferative syndrome→myeloproliferative disease and myeloproliferative disorder
  - forming of multi-word phrases: Coronavirus disease 2019
- Borrowing words from other languages: loan words are lexical borrowings adopted from foreign languages by root-for-root, word-for-word, or literal

translation. Some examples are chorion, diabetes, myopia, ophthalmia, pneumonia, trauma (Greek origin), femur, humerus, occiput, mandible, puncture, pulp (Latin origin), and diarrhea, diphtheria, disease, dislocation, malaise (French origin)

• Clipped words occur after discarding either the final (exam**ination**:পৰীক্ষা), beginning (**uni**versity:বশ্বিববদ্যিালয়), central (**influenza**: ফ্লু) or end part (polio**myelitis**: পলআি').

1.2.3 DIFFICULTIES OF MEDICAL TRANSLATION

Most people without a medical education do not understand the meaning of some typical sentences (text has its definition) written as Triage notes. The sentences do not have a subject due to those not being grammatically correct. Doctors annotate concisely by using a lot of jargon. These are the typical sense of the word but not considering a different language. NLP practitioners have the responsibility for cleaning these texts by using off-the-shelf NLP libraries and algorithms.

Consider the example of de-identified triage notes for deciphering taken from emergency room visits.

• States started last night, upper abd, took alka seltzer approx 0500, no relief. nausea no vomiting
• Since yesterday 10/10 constant Tylenol 1 hour ago. +nausea. diaphoretic. Mid abd radiates to back
• Generalized abd radiating to lower × 3 days accompanied by dark stools. Now with bloody stool this am. Denies dizzy, sob, fatigue.

Sentences have different semantics: "Sob" might be shortness of breath, different grammar: "since yesterday 10/10" (10/10 refers to the intensity of pain), and diverse vocabulary (abdominal can be abd).

1.3 APPROACHES TO BUILDING MT SYSTEM

MT solutions for the healthcare industry can broadly be categorized into rule-based, statistical, and neural types similar to usual MT categories. The evolution of MT is depicted in Table 1.3. Early Rule-based MT (RBMT) language experts manually

TABLE 1.3

Evolution of machine translation

Timeline	Evolution of MT
1950–1980	RBMT
1980–1990	EBMT
1990–2015	SMT
2015–Till Date	NMT

crafted many rules to translate one language into another language. It relies on sophisticated built-in linguistic rules, millions of bilingual dictionaries for each language pair, and extensive lexicons with morphological, syntactic, and semantic information. These rules were applied to the input text and generated the translation of the target texts. Users can improve translation quality by adding terminology into the translation process by creating user-defined dictionaries, which override the system's default settings. A quality RBMT system is computationally expensive due to training time and ongoing improvement of the system.

A significant development in MT happened in the 1990s when companies like IBM started to leverage statistical models that significantly improved translation quality. The corpus-based Statistical Machine Translation (SMT) approach to learning by automatically searching sentences and translating them into the target language. SMT searches for patterns in a large number of parallel texts able to assign the probability of a sentence from the target language being the translation of another sentence from the source language. Building an SMT system requires a massive number of parallel corpora between source and target languages at the sentence level. The quality of SMT extensively depends on the language pair of the specific domain being translated. The corpora building can often be challenging in the healthcare industry. There is a massive variation in named entities such as diseases, chemical compounds, active ingredients, gender, symptoms, dosage levels, dosage forms, route of administration, date, location, location-species, and adverse reaction. SMT technology is CPU intensive and requires an extensive hardware configuration to run translation models at a satisfactory performance quality. So, companies began to experiment with hybrid MT engines, which commonly combined SMT with RBMT. These advancements popularized MT technology and helped adoption on a global scale. The current state of the art in MT technology is Neural Machine Translation (NMT) harnesses the power of Artificial Intelligence (AI) and uses neural networks to generate translations. Language translation technology is continuously changing, bringing new functionalities and more significant benefits to the medical industry. The end-to-end training paradigm of NMT is the powerful modeling capacity of neural networks that can produce comparable or even better results than traditional MT systems. NMT uses a single large neural network to model the entire translation process, freeing the need for excessive feature engineering and employing continuous representations instead of discrete symbolic representations in SMT.

An encoder-decoder network is quite successful in different Recurrent Neural Network (RNN) variations in NMT consisting of two components: an encoder that consumes the input text and a decoder that generates the translated output text (Wolk and Marasek 2020). The encoder extracts a fixed-sized dense representation of the different length input texts. The task of the decoder is to generate the corresponding text in the destination language based on this dense representation from the encoder (Bahdanau et al. 2014; Cho et al. 2014). In 2017, MT made another technological breakthrough in NMT with the advent of transformer model (Vaswani et al. 2017) which is a state-of-the-art model for neural MT (Wu et al. 2016; Lakew et al. 2018; Wang et al. 2019). The model uses self-attention to speed up the training with significantly more parallelization. It follows the architecture of the

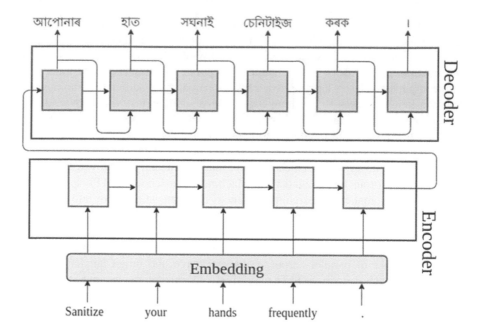

আপোনাৰ   হাত   সঘনাই   চেনিটাইজ   কৰক   ।

Sanitize   your   hands   frequently   .

FIGURE 1.1   Schematic diagram of medical-text translation.

encoder-decoder model by using stacked attention and point-wise fully connected layers for both the encoder and the decoder. The encoder can work on the input sequence in parallel and the decoder is auto-regressive. Previous output symbols influence each output and output symbols are generated one at a time (Figure 1.1). Several works in MT, especially in the health domain on low resource languages, are summarized in Table 1.4 below.

Since the application scenarios and markets for MT are extensive, many companies and organizations in different parts of the world are making attempts to build their own MT systems. Very few of them have focused on the medical and pharmaceutical industry, perhaps because of its tremendous difficulty in translation and high information security requirements. The professional translators convey the original tone and intent, considering cultural and regional differences between source and target languages. Professional medical translation services with the localization industry can help healthcare professionals make more informed decisions regarding treatments or medical procedures. As the English language has not been widely used for official communication in Assam, professional Assamese translation services can help them capture the healthcare industry. The trustworthy translation services (Integrated Language Solutions[2], Translation agency[3], PEC Attestation, Apostille & Translation Services India Pvt. Ltd.[4], Shakti Enterprise[5], Somya Translation Pvt. Limited[6], Linguainfo Language Translation Company[7], TRIDINDIA[8], and Honey Translation Services[9]) specifically related to medical and healthcare professionals supported by Assamese and other Indian languages save the lives of human beings.

TABLE 1.4

Major MT works on medical domain in low resource languages

Article	MT Approach	Evaluation/Key Findings
Zeng-Treitler et al. (2010)	RBMT	Babel Fish not adequate for medical records
Ruiz Costa-Jussà et al. (2011)	RBMT/SMT	SMT better than RBMT
Wu et al. (2011)	SMT	Best results of German, French, and Spanish
Dwivedi and Sukhadeve (2013)	RBMT	English-Hindi MT system for Homoeopathy.
Patil and Davies (2014)	SMT	GT not reliable for medical communication
Liu and Cai (2015)	Hybrid/SMT	GT performed better than hybrid system
Limsopatham and Collier (2015)	SMT	Map media messages to medical concepts
Wołk and Marasek (2015)	NMT	Comparison of SMT-NMT on Polish-English
Arcan and Buitelaar (2017)	NMT	Translating highly domain-specific expressions.
Khan et al. (2018)	NMT	Transfer learning by initializing parameters of NMT.
Skianis et al. (2020)	NMT	Generic, language-independent medical terminology translation

1.4 LEARNING TRANSPARENCY FOR PATIENT

It is essential to provide proper care to the patient. We can enhance the patient experience as it is always an epicenter of what technology evolved in the medical field by keeping patients' empathy in the heart, which is the best experience. Nowadays, people are more worried about their quality of care by moving to various doctor's chambers. The value-based care providers can create a care gap or performance report compared with a peer provider or government agency that enhances the value-based model. NLP can impact the patient experience and also value-based care. A patient can schedule the doctor's appointments online and prepare a form for the next day while staying away or sitting at home. In free-text clinical notes, much pertinent information for making correct predictions and recommendations is only available in healthcare. The free-text documents in the unstructured form are trapped enormously. NLP is a significant part of accumulating data from professional documents and clinical notes. The NLP is the primary use to transform the free (unstructured) text in documents and databases into normalized (structured) text. The structured data were suitable for analysis needed to make healthcare decisions that drive machine or deep learning algorithms. Horng et al. (2017) illustrate the benefit of extracting vital sign data and free text data to identify patients speculating of a life-threatening infection. These investigations used NLP to extricate data from the clinical text. Electronic health records (EHR) have become more prevalent across hospitals by implementing inpatient or ambulatory EHR systems.

Several MT systems for mobile or web applications facilitating doctor-patient communication have been built for low and under-resourced languages. Ahmad et al., (Musleh et al. 2016) developed a real-world Hindi-English SMT system for doctor-patient communication.

As healthcare advances to evolve to a more patient-oriented approach, patient expectations and demands will significantly push electronic communication. Many patients interested in using e-mail or other social media to communicate with their doctors are interested in receiving online health information from their doctor's office. For increasing numbers of providers and patients, Web messaging linked to a patient EHR is likely to become the preferred communication channel for routine clinical communications. A potential drawback of Web messaging is that it provides a less robust means of communication. Audio-video recordings allow patients to share information with caretakers and family members accurately. Experts agreed that one of the most significant benefits of recording visits is improving patients' recall and understanding of their medical conditions. Web messaging can be optimally integrated into healthcare delivery to improve safety, quality, and efficiency.

1.5 EVALUATION PROCEDURE AND METRIC ON MEDICAL DOMAIN

The evaluation of MT systems is important since its results show the degree of output reliability and are exploited for system improvements. Some freely available commercial software has implications of incorrect medical translation due to limitations in quality and considering ethnic diversity (Zeng-Treitler et al. 2010; Taylor et al. 2015; Anastasopoulos et al. 2021). There are many types of automated translation technology in the marketplace that can help automate the medical translation process. At times it becomes difficult for the translator to find exact words while translating because they have never heard of the new terms or esoteric expressions. An automated service was not a great fit for specific disaster vocabulary. The automatic translations hampered the Covid-19 response in some areas shown negatively impact communities of individuals with limited English proficiency during natural disasters. For example, in 2017, an automatic translation of a wildfire notice in California's Ventura county mistranslated the word "brush fire" using the Spanish word for "hairbrush" in place of "brush".

An inadequate translation in the medical field or misplacing of a word by a nonprofessional can lead to tragedy and cost a massive amount in medical malpractice compensation. In 2007, 47 patients had gone through a second knee replacement operation due to inaccurate translation. The translator translated the phrase "non-modular cemented" as "without cement" or "non-cemented", resulting in painful methods that needed months of recovery. In 1980 at Florida hospital, Willie Ramirez ended up quadriplegic because a certified medical interpreter in Oregon translated the Spanish word "intoxicado" as "intoxicated" which means "ingested something" (Spanish). A slight mistranslation can lead the healthcare practitioner down the wrong path, although they realize that a mistake was made, which may be difficult to backtrack in later stages.

Unlike general translations, medical translations should be done by highly-qualified translators who possess tremendous knowledge in the specific field. The quality of the text depends on sound medical knowledge and personal interest in the text.

Many anatomical and clinical terminologies that persist in medicine today are Latin or Latinized Greek words. These words can be traced back to the golden age of Greek civilization during the fifth century BC. Microsoft announced that Microsoft Translator would help users translate conversations, street signs, websites, and documents to Assamese language and vice-versa.

Evaluating the quality of MT requires an automatic method as human evaluation would be highly time-consuming and cost-inefficient to be evaluated. The criteria of translation quality are its adequacy and fluency. The most common evaluation metric is BiLingual Evaluation Understudy (BLEU) (Papineni et al. 2002). Given a human-generated reference sentence with a corresponding translated sentence, it calculates a score by comparing *n-gram* overlap. Two separate translators can not produce identical translations for the same sentence in the same language pair. A set of several rounds of iterations are required for fulfilling the client's requirement. Automated translations find difficulties in interpreting contextual and cultural elements of a text and quality is dependent on the type of system and how it has been trained.

The adequacy, fluency measures the effectiveness of a translation. Adequacy expresses meaning from the source language to the target language. Fluency measures the grammatically well-formed and ease of interpretation of the sentence (idiomatic word choices). Different word choices and changing the word order that conveys the same meaning is the challenge of evaluating translations for a sentence.

BLEU, a corpus-based metric, calculates the automatic quality score for MT systems that estimate the correspondence between a translated output and a human reference translation (Papineni et al. 2002). The primary notion of BLEU is closer to a professional translation with its machine-translated output. BLEU counts the number of matches by comparing the *n-gram* of the candidate translation with the *n-gram* of the reference translation. The more matches, the better the translation quality, where matches are independent of their positions.

A low BLEU score means a high mismatch. A sequence of words or tokens occurring within a given window (where n is window size) is known as n-gram. A perfect match and mismatch result in a score of **1.0** and **0.0**, respectively. A translation that possesses exact words as in the references (more than one reference translation) satisfies the adequacy. The longer n-gram matches between reference and candidate translation tend to capture fluency. A BLEU score runs on a scale from **0** to **1**. The score is expressed as percentages rather than decimals (turned into a **0** to **100** scale) for better readability.

Mathematically, the BLEU score formula consists of two parts: the brevity penalty (BP) and the n-gram overlap are shown in the following Equation (1.1).

$$BLEU = BP . \, exp \sum_{n=1}^{N} w_n log p_n \qquad (1.1)$$

where $BP = \{ 1, \, if \, c > r e^{\left(1-\frac{r}{c}\right)}, \, if \, c \leq r$ and $P_n = \frac{\sum \sum Count \, (n-gram)}{\sum \sum Count \, (n-gram)}$

The brevity penalty (BP) penalizes the BLEU score if the candidate sentence is shorter than the reference sentence. BP compensates for the possibility of high precision translation that is too short. Here **c** is the total number of unigrams

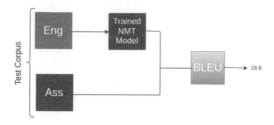

FIGURE 1.2   Schematic diagram of evaluation.

(length) in all the candidate sentences, and **r** is the sum of effective reference sentence length for each candidate sentence in the corpus. The two essential characteristics of translation are adequacy and fluency captured by the modified n-gram precision p_n score. w_n is n-gram precision weight. BLEU uses the value of **N** as 4. The number of word count for each candidate to its associated maximum reference count is a clip. Figure 1.2 depicts an evaluation process of the MT system using an automated evaluation matric BLUE. The test corpus of English(Eng)-Assamese(Ass) shows a dummy score of 28.8 for English to Assamese translation.

An example of reference sentence and its several translated (candidate) outputs from English (source) to Assamese (target) are illustrated here. The n-gram match (length) between candidate translation and the given target reference is shown at the end of each candidate translation.

> **Source Text:** Wearing a face mask is compulsory in all public places.
> **Target Text:** সকলো ৰাজহুৱা স্থানত মুখা পনিধাটো বাধ্যতামূলক ।
> **Candidate1:** মুখা পৰিধান কৰাটো এক বাধ্যতামূলক নিয়ম । (2-gram)
> **Candidate2:** মুখৰ মুখা পনিধাটো বাধ্যতামূলক । (3-gram)
> **Candidate3:** ৰাজহুৱা ঠাইত মুখা পনিধাটো বাধ্যতামূলক কৰা হৈছে । (4-gram)
> **Candidate4:** সকলো ৰাজহুৱা স্থানত মুখৰ মুখা পনিধাটো বাধ্যতামূলক । (5-gram)

Nowadays, close derivatives of BLEU (METEOR, NIST, LEPOR, and F-Measure) are often used to compare the quality of different MT systems in enterprise use settings. Researchers worked on evaluating medical terminologies by SacreBLEU, BLEU, METEOR, and TER metrics as well. Skianis et al. (2020), attempted to develop a first baseline translation from English to French on numerous medical terminologies and datasets leveraging SMT and NMT present promising results for the (International Classification of Diseases) ICD11 classification (Skianis et al. 2020).

1.6   CURRENT TECHNOLOGIES

This section discusses the current state-of-the-art of MT technologies in clinical practices, healthcare, and medicine. Further, we investigate additional recent innovations in technology that can leverage the language industry within healthcare.

In 2010, deep neural network-style machine learning methods became widespread in natural language processing and achieved state-of-the-art results in many tasks. It is increasingly important in healthcare and medicine, where NLP is being used to analyze text and notes in electronic health records.

Different state-of-the-art deep learning techniques such as healthcare-specific named entity recognition models, word embeddings, and entity resolution models can extract clinical data from text. The improved methods of collecting high-quality data and advancements in the machine (deep) learning models fueled a new wave of healthcare practices. An EHR essentially stores patient records in unstructured and structured formats. To create a more intelligent healthcare system in which the best treatment decisions are computationally learned from electronic health record data by deep-learning methodologies. International technology giants like Google, Microsoft, IBM, and Amazon are all keen on developing MT. Since Google launched the NMT system in 2016, the improvement of the quality of MT has achieved more attention and interest from all other NLP tasks.

The transformer is a new type of neural network model that emerged in 2017 based on self-attention. The transformer model replaced previously dominated RNN (its variations are LSTM/GRU) and become a state-of-the model in MT and many other NLP tasks. Compared to RNN, transformers have much higher computational efficiency and can efficiently exploit the modern parallel hardware (GPU/TPU). It allows training on much larger models on a massive amount of data. The transformer framework overcomes the bottleneck in which the recurrent neural network model cannot be calculated in parallel. Many NMT engines produce sporadic errors while training the system. The popular approach in deep learning is pre-trained models which have been previously trained on large datasets. A pre trained transformer model by fine-tuning can further improve performance, requiring fewer data and computational resources. It might help in the medical translation in low-resource languages.

The highly lexicalized nature of languages causes sensitivity of domain shift in the NMT system. One solution is lexicon induction to obtain an in-domain lexicon and construct a pseudo-parallel in-domain corpus. The in-domain monolingual target corpus use word-by-word back translation for constructing the synthetic parallel in-domain corpus (Edunov et al. 2018; Hu et al. 2019). Furthermore, applying a pseudo-in-domain corpus with fine-tuning, a pre-trained out-of-domain NMT model called the unsupervised adaptation method is another explanation. Domain adaptation in MT can be applied when a large amount of out-of-domain data co-occur with a small amount of in-domain data (Soares and Becker 2018). A domain adaptation experiment containing a medical domain with the lexicon inclusion performs an acceptable accuracy in low resource language NMT. NMT with automated customization using domain-specific corpora say the medical domain is known as domain adaptation in MT (Arcan and Buitelaar 2017).

Intento[10] evaluated six domain adaptive NMT systems for English-to-German translation using biomedical corpora[11] of several sizes (from 10K to 1M segments) and evaluated them compared to stock MT engines. It was a breakthrough moment in MT, probably the biggest one since the invention of NMT.

A word in a sentence is often related to multiple domains that indicate its domain preference. The word in a distributed representation possesses embedding by mixing domain proportions from different domains. In a transformer architecture for different domains, carefully designing dot-product multi-head attention modules can achieve effective domain knowledge sharing in multi-domain NMT (Zeng et al. 2018).

A technique most commonly used in NMT, especially in a low resource language, is transfer learning, which falls under domain adaptation. Transfer learning is the process where a child model in one language pair (in-domain data) is trained by transferring the knowledge learned from an existing parent model in another language pair (out-of-domain data). By initializing the parameters in NMT from the previous model, an increase in training accuracy on out-of-domain and multiple in-domain datasets has been achieved for biomedical corpora (Khan et al. 2018; Peng et al. 2019).

Advanced models are used to predict hospital and professional billing codes for administrative cost reduction and billing process improvements using deep learning techniques (Joo et al. 2021). The rare words can not be translated correctly by a conventional NMT system. These are called out-of-vocabulary (OOV). NMT replaces these OOV as <unk> tokens that do not have any information. <unk> (unknown words) are unique words that cannot be translated into the target token during generation. So the possibility of a loss of translated information. A lexicon of biomedical vocabulary, MedDRA (Bo et al. 2007), is used for semantic disambiguation model to solve <unk> problem (Liu et al. 2020).

The roadblocks to bringing medicine into the data-driven period are cultural and operational. It is time to safely bring huge medical data repositories and advanced learning algorithms together with physicians to make a deep-learning healthcare system. Deep learning, the newest iteration of machine learning methodologies, is now performing at state-of-the-art levels in previously difficult tasks such as language processing, information retrieval, and forecasting. India is home to many native languages that become linguistic diversity in a multilingual country with a growing population. Multilingual challenge with multilingual technology. A study from Google (Johnson et al. 2017) showed that using multilingual data when training NMT systems can improve translation performance, especially when using a many-to-one scheme.

The Covid-19 pandemic has shown the need for multilingual access to hygiene and safety guidelines and policies (Zeng-Treitler et al. 2010). A Multilingual Neural Machine Translation (MNMT) can be employed to translate biomedical text (O'Brien and Federici 2019). A large number of domain tags from generic and biomedical data use to train the MNMT system (Bérard et al. 2020).

1.7 CONCLUSION

Medical translation is a crucial factor in disseminating new knowledge and discoveries in the medical field. Still, it can also be a critical factor in the provision of global and foreign health services.

It cannot deny the need for machine-translated content in healthcare. Its credibility and increased use of social media during Covid-19 pandemic can only be expedited with more robust training data for learning the model (O'Neill 1998).

The traditional approaches for translating text are time-consuming and require a series of labor-intensive levels. Because of the inefficiency and expense of conventional translation, companies and agencies are looking for faster, cost-efficient, and better performance in terms of accuracy. The neural MT with transformer model has been using a state-of-the-art method and architecture for MT. Language technologies have been steadily advancing to deliver high-quality translated documents that can be used for official purposes.

Covid-19 is posed the history's biggest translation challenge because it not only involved translating one or a small number of primary languages in a single region but also on a scale of thousands of languages across the world. Several crowd-sourced translation projects spread the knowledge to fight Covid-19 to healthcare workers worldwide[12].

In medicine, there is no room for mistakes and errors, which is a challenge for translators. Different language technologies and advanced methods can overcome that. In the not-too-distant future, MT will be capable of translating a text in the biomedical domain at the required quality. It will improve patient monitoring, which will improve patient outcomes. Now, the machine is acting as the rescuer in medicine in day-to-day life.

NOTES

1 https://censusindia.gov.in/2011Census/C-16_25062018_NEW.pdf
2 https://www.integratedlanguages.com/
3 https://translation.agency/
4 https://pecattestation.com/
5 http://www.shaktienterprise.com/
6 http://www.somyatrans.com/
7 https://linguainfo.com/
8 https://www.tridindia.com/
9 http://www.honcytranslations.com/
10 https://inten.to
11 https://ufal.mff.cuni.cz/ufal_medical_corpus
12 https://covidtranslate.org

REFERENCES

Anastasopoulos, A., L. Besacier, J. Cross, M. Gallé, P. Koehn, and V. Nikoulina. 2021. On the evaluation of machine translation for terminology consistency. arXiv preprint arXiv:2106.11891

Arcan, M., and P. Buitelaar. 2017. Translating domain-specific expressions in knowledge bases with neural machine translation. CoRR. arXiv preprint arXiv:1709.02184

Bahdanau, D., K. Cho, and Y. Bengio. 2014. Neural machine translation by jointly learning to align and translate. arXiv preprint arXiv:1409.0473

Bérard, A., Z. M. Kim, V. Nikoulina, E. L. Park, and M. Gallé. 2020, December. A multilingual neural machine translation model for biomedical data. In Proceedings of the 1st Workshop on NLP for Covid-19 (Part 2) at EMNLP 2020.

Bo, Q. Y., N. N. Xiong, J. D. Zou, M. Jiang, F. Liu, and Z. W. Anna. 2007. Internationally agreed medical terminology: Medical dictionary for regulatory activities. Chinese Journal of Clinical Pharmacology and Therapeutics, 12, no. 5: 586.

Chabner, D. E. 2020. The Language of Medicine E-Book. Elsevier Health Sciences.

Cho, K., B. van Merriënboer, C. Gulcehre, D. Bahdanau, F. Bougares, H. Schwenk, and Y. Bengio. 2014, October. Learning phrase representations using RNN encoder–decoder for statistical machine translation. In Proceedings of the 2014 Conference on Empirical Methods in Natural Language Processing (EMNLP) (pp. 1724–1734).

Dwivedi, S. K., and P. P. Sukhadeve. 2013, August. Comparative structure of Homoeopathy language with other medical languages in machine translation system. In 2013 International Conference on Advances in Computing, Communications and Informatics (ICACCI) (pp. 775–778). IEEE.

DžUGANOVá, B. 2013. English medical terminology–different ways of forming medical terms. *Jahr: Europski časopis za bioetiku*, 4, no. 1: 55–69.

Edunov, S., M. Ott, M. Auli, and D. Grangier. 2018. Understanding back-translation at scale. In Proceedings of the 2018 Conference on Empirical Methods in Natural Language Processing (pp. 489–500).

Fischbach, H. (Ed.). 1998. *Translation and Medicine*. John Benjamins Publishing.

Hu, J., M. Xia, G. Neubig, and J. G. Carbonell. 2019, July. Domain adaptation of neural machine translation by Lexicon induction. In Proceedings of the 57th Annual Meeting of the Association for Computational Linguistics (pp. 2989–3001).

Johnson, M., M. Schuster, Q. Le, M. Krikun, Y. Wu, Z. Chen, ... and J. Dean. 2017. Google's multilingual neural machine translation system: Enabling zero-shot translation. *Transactions of the Association for Computational Linguistics*, 5: 339–351.

Joo, H., M. Burns, S. S. K. Lakshmanan, Y. Hu, and V. V. Vydiswaran. 2021. Neural machine translation–based automated current procedural terminology classification system using procedure text: Development and validation study. *JMIR Formative Research*, 5, no. 5: e22461.

Karwacka, W. 2015. Medical translation. *Ways to Translation*, 271–298.

Khan, A., S. Panda, J. Xu, and L. Flokas. 2018, October. Hunter nmt system for wmt18 biomedical translation task: Transfer learning in neural machine translation. In Proceedings of the Third Conference on Machine Translation: Shared Task Papers (pp. 655–661).

Lakew, S. M., M. Cettolo, and M. Federico. 2018, August. A comparison of transformer and recurrent neural networks on multilingual neural machine translation. In Proceedings of the 27th International Conference on Computational Linguistics (pp. 641–652).

Lewis, W. 2010, May. Haitian Creole: How to build and ship an MT engine from scratch in 4 days, 17 hours, & 30 minutes. In 14th Annual conference of the European Association for Machine Translation.

Limsopatham, N., and N. Collier. 2015, September. Adapting phrase-based machine translation to normalise medical terms in social media messages. In Proceedings of the 2015 Conference on Empirical Methods in Natural Language Processing (pp. 1675–1680).

Liu, W., and S. Cai. 2015, July. Translating electronic health record notes from English to Spanish: A preliminary study. In Proceedings of BioNLP 15 (pp. 134–140).

Liu, H., Y. Liang, L. Wang, X. Feng, and R. Guan. 2020. BioNMT: A Biomedical neural machine translation system. *International Journal of Computers, Communications & Control*, 15, no. 6.

Musleh, A., N. Durrani, I. Temnikova, P. Nakov, S. Vogel, and O. Alsaad. 2016, April. Enabling medical translation for low-resource languages. In International Conference on Intelligent Text Processing and Computational Linguistics (pp. 3–16). Springer, Cham.

O'Brien, S., and F. M. Federici. 2019. Crisis translation: Considering language needs in multilingual disaster settings. *Disaster Prevention and Management: An International Journal*.

O'Neill, M. 1998. Who makes a better medical translator: The medically knowledgeable linguist or the linguistically knowledgeable medical professional? A physician's perspective. *Translation and Medicine*, 194.

Papineni, K., S. Roukos, T. Ward, and W. J. Zhu. 2002, July. Bleu: A method for automatic evaluation of machine translation. In Proceedings of the 40th annual meeting of the Association for Computational Linguistics (pp. 311–318).

Patil, S., and P. Davies. 2014. Use of google translate in medical communication: Evaluation of accuracy. *BMJ (Clinical Research ed.)*, 349: g7392

Peng, W., J. Liu, L. Li, and Q. Liu. 2019, August. Huawei's NMT systems for the WMT 2019 biomedical translation task. In Proceedings of the Fourth Conference on Machine Translation (Volume 3: Shared Task Papers, Day 2) (pp. 164–168).

Rogl, R. 2017. Language-related disaster relief in Haiti. *Non-Professional Interpreting and Translation: State of the Art and Future of an Emerging Field of Research*, 231–255.

Ruiz Costa-Jussà, M., M. Farrús Cabeceran, J. B. Mariño Acebal, and J. A. Rodríguez Fonol-Losa. 2011. Automatic and human evaluation study of a rule-based and a statistical Catalan-Spanish machine translation systems. In Seventh Conference on International Language Resources and Evaluation (pp. 1707–1711).

Skianis, K., Y. Briand, and F. Desgrippes. 2020, November. Evaluation of machine translation methods applied to medical terminologies. In Proceedings of the 11th International Workshop on Health Text Mining and Information Analysis (pp. 59–69).

Soares, F., and K. Becker. 2018, October. UFRGS participation on the WMT biomedical translation shared task. In Proceedings of the Third Conference on Machine Translation: Shared Task Papers (pp. 662–666).

Taylor, R. M., N. Crichton, B. Moult, and F. Gibson. 2015. A prospective observational study of machine translation software to overcome the challenge of including ethnic diversity in healthcare research. *Nursing Open*, 2, no. 1: 14–23.

Vaswani, A., N. Shazeer, N. Parmar, J. Uszkoreit, L. Jones, and A. N. Gomez. 2017. Ukasz Kaiser, and Illia Polosukhin. 2017. Attention is all you need. *Advances in Neural Information Processing Systems*, 5998–6008.

Wang, Q., B. Li, T. Xiao, J. Zhu, C. Li, D. F. Wong, and L. S. Chao. 2019, July. Learning deep transformer models for machine translation. In Proceedings of the 57th Annual Meeting of the Association for Computational Linguistics (pp. 1810–1822).

Wołk, K., and K. Marasek. 2015. Neural-based machine translation for medical text domain. Based on European Medicines Agency leaflet texts. *Procedia Computer Science*, 64: 2–9.

Wolk, K., and K. P. Marasek. 2020. Translation of medical texts using neural networks. In *Deep Learning and Neural Networks: Concepts, Methodologies, Tools, and Applications* (pp. 1137–1154). IGI Global.

Wu, Y., M. Schuster, Z. Chen, Q. V. Le, M. Norouzi, W. Macherey, ... and J. Dean. 2016. Google's neural machine translation system: Bridging the gap between human and machine translation. arXiv preprint arXiv:1609.08144

Wu, C., F. Xia, L. Deleger, and I. Solti. 2011. Statistical machine translation for biomedical text: Are we there yet?. In *AMIA Annual Symposium Proceedings* (Vol. 2011, p. 1290). American Medical Informatics Association.

Zeng, J., J. Su, H. Wen, Y. Liu, J. Xie, Yin, Y., and Zhao, J. 2018. Multi-domain neural machine translation with word-level domain context discrimination. In Proceedings of the 2018 Conference on Empirical Methods in Natural Language Processing (pp. 447–457).

Zeng-Treitler, Q., H. Kim, G. Rosemblat, and A. Keselman. 2010. Can multilingual machine translation help make medical record content more comprehensible to patients?. In *MEDINFO 2010* (pp. 73–77). IOS Press.

2 Feature Analysis and Classification of Impaired Language Caused by Brain Injury

Marisol Roldan-Palacios and Aurelio López-Lopez
INAOE- Instituto Nacional de Astrofísica Óptica y
Electrónica, Puebla, México

CONTENTS

2.1 INTRODUCTION

Linguistic capabilities can be slightly or severely affected by traumatic brain injury (TBI) incidents. This communication impairment can encompass different aspects as those complications in production or comprehension of language presented in aphasia or those difficulties in the articulation of speech present in dysarthria, both possibly present in TBI-affected language. Depending on the severity of the deficit, social integration might be deteriorated, affecting in consequence the quality of life.

DOI: 10.1201/9781003138013-2

19

The study of these kinds of languages is complicated given the relative scarcity of the data, with the additional problem of the impossibility of generating synthetic samples. These factors lead to consider this type of language manifestation a low-resource language.

This chapter details efforts to reveal modified patterns of TBI-affected language, focused on the analysis of the responses of components characterizing such altered language as a consequence of a TBI, but especially during the post-traumatic period, which so far has been barely analyzed. The purpose of the work is oriented to provide elements to support speech-language pathologist decision-making in the accompanying during the recovery therapy period. For this, our main aim is to reveal which features contribute the most information, while they are sensitive enough to reflect subtle changes in the response in TBI-affected language, across the registered recovery time points.

2.2   PREVIOUS RESEARCH

Language impairment has been extensively investigated from the linguistic point of view. With a variety of causes, the range and spectrum of affectations are wider, implying the need for many additional efforts to understand their development and nature, mainly when they have an inherent complexity. However, possibly because of the extensive established criteria and sometimes with differences in the employed notions, together with the absence of a standardized protocol to elicit samples, they often reached divergent findings. In that respect, there are steady efforts to provide and unify proper mechanisms for an assertive condition assessment (MacWhinney 2000, 2020), developed in different directions, one of them working in gathering a TBI Corpus (Elbourn et al. 2018, Stubbs et al. 2018).

A Statistical Language Model was previously proposed, defining the n-grams of Part-Of-Speech (POS) tags (Solorio and Yamg 2008) or directly applied to produced words (Pakhomov et al. 2010), to approximate probabilities to determine their models. Focusing attention on the appropriateness of the components involved in the performed assessment, an ACT-R architecture was used as the basis for establishing a knowledge model (Oliva et al. 2014). Natural Language Processing (NLP) techniques were introduced by (Peintner et al. 2008), basically to discriminate between the instances belonging to one of the study groups or the negative cases (Gabani et al. 2009, Fraser et al. 2013, Rentoumi et al. 2017). Statistical language model 'Perplexity' was employed, this time as a component (Gabani et al. 2009), in addition to context-free grammar (CFG) production rules (Fraser et al. 2014a), using ablation to complement attribute filtering (Fraser et al. 2014b). Furthermore, Leave-One-Out Cross-Validation (LOOCV) (Rentoumi et al. 2017) was considered to handle overfitting, and synthetic instances were also generated (with SMOTE), aiming to test the implemented method.

Among the explored learning algorithms, we found the following: Simple Logistic, J48 decision tree, Multilayered perceptron (neural network) (Peintner et al. 2008), Naive Bayes (Gabani et al. 2009, Rentoumi et al. 2017), Support Vector Machine (Gabani et al. 2009, Fraser et al. 2014a, 2014b), Random Forest (Fraser et al. 2014b), Artificial Neural Networks (Gabani et al. 2009), Boosting with Decision Stump (Gabani et al. 2009), and Sequential Minimal Optimization (Rentoumi et al. 2017).

Approaches with and without transcriptions have both benefits and drawbacks (Steel and Togher 2018). While the latter requires less time investment, the former tends to be error-free which represents solid grounds to start any language analysis, mainly when its deficiencies are the subject of the study. In the recent statistics works conveying research on TBI-language deficits, they were carried out on samples collected in the first post-injury recovery year and working on different tasks. For instance, for transcriptions of narrative discourse, the main concept analysis (MCA) together with a coherence analysis (Young 2017) were the basis to report group changes on these indices over three time-points: 3–6, 12, and 24 months. As a primary measure, MCA was also explored in combination with variables defined as pre-injured, injured, and post-injured (Elbourn et al. 2018), to generate a mixed linear model of individual recovery for participants at 3, 6, 9, and 12 months after the incident. Measures of productivity along with informativeness and story organization (Power ct al. 2020) or combined with macrostructural level characteristics based on a checklist (Stubbs et al. 2018) were extracted to inform recovery performance at two time-points: 3 and 6 months after the injury.

2.3 DATA DESCRIPTION

Based on a long-term study (Elbourn et al. 2018, Stubbs et al. 2018), a number of samples were collected post-injury at periodic stages from a group of participants for two years. The assembled Corpus has become part of TBIBank (TBIBank: 2001) as a contribution to the investigation. For the group of interest, we have available examples for 3, 6, 9, 12, and 24 months, although the latter sample has been partially transcribed, and the current proportion is 1:5 in comparison to the size of the negative group. The population age ranges from 17 to 66 years old. Discourse transcription related to the 'Cinderella' retelling story task were the samples used in our study. For the negative cases, the corresponding age range is from 16 to 63 years old, and the task is the generation of a story based on the 'The Runaway' picture. These samples come from a different study (Coelho et al. 2005) also exploring language modification in people after suffering a TBI but years later of their incidents.

2.4 FEATURE SETS AND ANALYSIS

For our chapter study, two feature sets were included. The first is a part of the computed NNLA (Northwestern Narrative Language Analysis) (Thompson 2013) (MacWhinney 2000, 2020), here on referred to as CNNLA. The second set is a cluster of indices regulated by directives reflecting the index of syntax productivity (Scarborough 1990, MacWhinney 2000, 2020), here referred to as IPSYN feature set. These sets of attributes were extracted employing CLAN (Computerized Language Analysis) (MacWhinney 2000, 2020, TalkBank 2021).

CNNLA. This set consists of a collection of 25 lexical features composed of part-of-speech (POS) densities as well as their proportions and ratios, such as words per minute, the named open-class words referencing to (for example to all nouns and verbs no auxiliaries), number of auxiliaries (*aux*), number of modals and modal auxiliaries (*modals*), the percentage of close-words with respect to the total words into the instance in turn, as well as the utterance by minute (*UttsByMin*).

IPSYN. This set is subdivided into four groups, *noun phrase*, *verb phrase*, *question negation*, and *sentential structure*. For instance, in the noun phrase, the attribute called *N3* consider modifiers as adjectives, possessives, and quantifiers, and *N1* proper, mass or count noun to earn their first point. If there are second occurrences of these same types and the first and second stems are not equal, then a second point is added. The characteristic named *V3*, belonging to the verb phrase set has a point if the combination *prep+$NP* wherein $NP we find expressions as *article+modifier+noun*.

2.5   EXPERIMENTS

A correlation analysis was carried out over the CNNLA feature set; however, the same analysis was not very indicative on IPSYN feature sets, in consequence, they were inspected differently. The characteristics were removed until the rest had a null or quite poor relation among them. The evolution course of the remaining lexical and syntactical components was recorded both, as a whole and individually, to analyze and determine their response across the five stages of recovery. Regarding the instance's aggregation process, except for the 24-recovery point, which is smaller, in general, the size of the samples moves in an average of 85 elements, in a relatively balanced combination of negative and study group elements. One last consideration concerning the last recovery point, corresponding to 24 months, is that this was extended with samples taken at such recovery stage but representing the transcriptions for the generative-story task, based on pictures, also part of the same corpus, to bring the obtained responses near to the real picture. Details are discussed next.

2.5.1   EXPERIMENTAL SETTING

As pre-processing steps after extracting the attributes, when reviewing the negative group, four instances were removed. Furthermore, a normalization along with a ten-factor scaling was applied to the 25-CNNLA set. Next, the feature set was divided into groups reasonably related:

L_1 = {'DurationSec', 'WordsByMin', 'TotalUtts', 'TotalWords', 'MLUWords', 'UttsByMin'},
L_2 = {'openClass', 'PerOpenClassByWords', 'closedClass', 'PerClosedClassByWords', 'PerOpenClosed'},
L_3 = {'Nouns', 'PerNounsByWords', 'nounByVerb'},
L_4 = {'Verbs', 'PerVerbsByWords', 'nounByVerb'},
L_5 = {'adj', 'adv', 'det', 'pro', 'aux', 'conj', 'complementizers', 'modals', 'prep'}.

Then, internally, the correlation among the components of each subset was evaluated to identify those nulled or minimally correlated. The angle of separation among the vectors representing each attribute, along with the correlation matrix, guided the selection of the features considered by pairs in each subset. Afterward, the same technique was replicated across the rest of the subsets. The IPSYN feature set, as stated above, consists of subsets, $N = \{N_1, ..., N_{11}\}$, $V = \{V_1, ..., V_{16}\}$, $Q = \{Q_1, ..., Q_{11}\}$, $S = \{S_1, ..., S_{20}\}$, based on them, the correlation analysis was performed, not before a preliminary filter, removing those directives representing simultaneously a rule and a 'must' condition for other. However, due to the values domain in which IPSYN rules move, correlation results could not be double-checked as in the CNNLA feature set, then we decided to proceed with the analysis on the subsets after removing condition rules.

2.5.1.1 Feature Engineering Steps

The steps to carry out the feature selection task are summarized in the following processes.

2.5.1.1.1 CNNLA Feature Selection

Input: Feature Set

- Normalization
- Initial Segmentation
- For all subsets do
  - Correlation analysis.
  - Remove redundancy.
  - Feature subset assessment with LM.

2.5.1.1.2 IPSYN Feature Selection

Input: Feature set

- For all rule subsets do
  - Remove internal dependencies.
  - Individual feature assessment with LM.

2.5.1.2 Selected Features

After completing those steps, the resulting clusters were:

{'TotalWords', 'MLUWords', 'UttsByMin'},
{'PerOpenClosed'},
{'nounByVerb'},
{'aux','complementizers', 'modals', 'prep'}

corresponding to the lexical part, and $N_R = NP/\{N_4\}$, $V_R = VP/\{V_1, V_2, V_5, V_6, V_8, V_9\}$, $Q_R = QN/\{Q_1, Q_2, Q_3, Q_4, Q_5\}$, $S_R = SS/\{S_1, S_2, S_3, S_5, S_6, S_8, S_{11}\}$ concentrate the responses of the original component subsets. Where $NP = \{N_i\}$ with $i = \{1, ..., 11\}$, $VP = \{V_j\}$ with $j = \{1, ..., 16\}$, $QN = \{Q_k\}$ with $k = \{1, ..., 11\}$, and $SS = \{S_t\}$ with $t = \{1, ..., 20\}$ denote the *noun phrase, verb phrase, question-negation, sentence structure* policies sets. V_7 and V_{12} were not removed from the VP set because they behave as a constraint in the SS subset, but not inside their feature group.

Based on individual evaluations, an IPSYN subset was determined, leaving the most contributed element for NP, VP, and SS collections, QN group was excluded since it produced scarce information as a whole and in individual assessment. Then the subset is defined as $\{N_1, V_3, S_{17}\}$. Because of the diverse but natural relationship that those components maintain, we hypothesized that they will be in harmony, having as a result, a trajectory grade on top of single weighting.

2.5.1.3 Evaluation

We performed a supervised discrimination task with an average of ±42 instances per group, considering study and negative instances. A leave-one-out cross-validation was applied in all experiments across wideness and deepness, i.e., to the whole feature clusters and to the whole stages of recovery analyzed.

2.5.2 RESULTS FOR LEXICAL ANALYSIS

The responses of each component along with their efficacy as a set are presented in graphics to observe how their behavior varies during the studied recovery stages, as well as to be sensitive to how they evolve in comparison with each other.

Decision tree learning methods have been applied before in similar or smaller data sets (Peintner et al. 2008, Fraser et al. 2014b). These are not particularly demanding on data. For comparison, we also included one of the applied probabilistic learning algorithms.

So, we employed a variety of learning algorithms for the assessment step after smaller subsets were determined, though only three of them are reported here, Random Forest (RF), A Fast Decision Tree Algorithm (SPAARC), and Bayes Net. Figures 2.1, 2.2, 2.3 contain the curves progressing on the stages of recovery (the X-axis), against the accuracy achieved on the Y-axis, where each curve

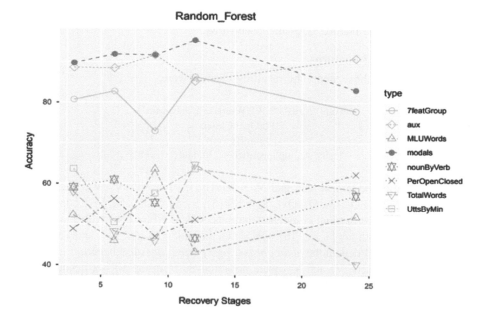

FIGURE 2.1   Random forest accuracy, seven CNNLA-feature set.

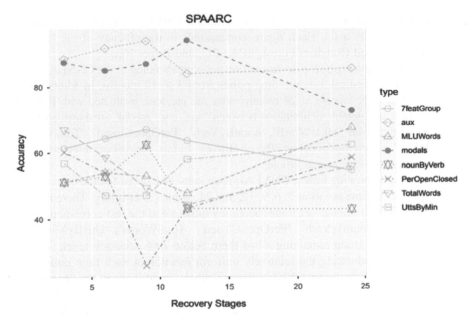

FIGURE 2.2   SPAARC accuracy, seven CNNLA-feature set.

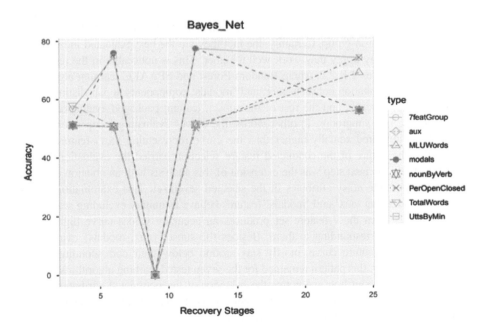

FIGURE 2.3   Bayes net accuracy, seven CNNLA-feature set.

corresponds to the evaluation of each determined CNNLA-feature, in addition to their assessment as a set. Each figure corresponds to the efficacy of the learning algorithms reported (Random Forest, SPAARC, and Bayes Net).

A line chart showing how a seven-features set corresponding to the CNNLA-group, react over the periodical 3-month time points from 3 to 12 months, in addition to a final recovery stage sample at 24 months after the incident, evaluated with Random Forest. There is a notable delimitation between 'aux' and 'modal' attributes accuracy levels remaining over 'MLUWords', 'nounByVerb', 'PerOpenClosed', 'TotalWords', 'UttsByMin', which moves below 70% of efficacy. In fact, the pair of observed curves maintain over the response as whole.

A line chart showing how a CNNLA-group subset of seven features, react each one over a set of time points at 3, 6, 9, 12, and 24 months after injury. There is a gap between 'aux' and 'modal' attributes accuracy indices and the rest corresponding to 'MLUWords', 'nounByVerb', 'PerOpenClosed', 'TotalWords', 'UttsByMin', and the evaluation in group remaining all of them below 70% accuracy level.

A line chart exhibiting the relatively uniform reaction at each time point from 3 to 24 months of the seven-CNNLA feature subset, in addition to the change as a whole. In general, moving between 50% to 80%, having a drastic fall at 9 months where zero is reached, to then going up until the levels are mostly handled.

2.5.3 ANALYSIS AND DISCUSSION OF RESULTS FOR LEXICAL ANALYSIS

A correlation analysis was performed on the features to determine those less related to identify the more representative set, under the conjecture that they or part of them could contribute the most in-group. Certainly, the fact they are the best evaluated individually does not necessarily imply they work well together. This is noticeable in the responses illustrated in Figures 2.1, 2.2, where Random Forest and SPAARC learning algorithms show a clear delimitation between 'aux' and 'modals' components, a pair distinguished by the highest efficacy, and the rest of the subset, leaving packaged evaluation in the middle of the two subgroups. In other words, the curves evaluating independently aux and modals appeared actually higher than the curve representing the 7-feature set. It seems that the nature of the conducted feature selection carried the evaluation to this position. Then, the next step was the extension of this analysis with an ablation exercise to identify possible noisy attributes in the selected seven-set. This examination corroborated the idea that 'aux' and 'modals' features behave as unitary excluding sets. None of the subsets from the 7-feature set produces an accuracy values curve higher than the trajectories corresponding to them. Besides the subset {aux, modals} evaluations surpassing the 7-feature curve, mostly stay around below their corresponding single weights. Moreover, this pattern remained for the seven tested learning algorithms. So far, the observations about how we notice the behavior of the nominated attributes, in the following, the consideration of how they correspond to the alteration in TBI-affected language reaction through the post-traumatic recovery stages studied are presented.

First, it is necessary to clarify that a lower obtained efficacy is interpreted as progress in the TBI-affected language status. After responses examination of each component, according to the results of the study, it was not possible to perceive a clear improvement or decline in the examined language. The preceding is remarked because the efficacy recorded percentage varies up and down as the recovery stages move forward. Though, for instance, the 7-feature set SPAARC evaluation shows progress relatively stable at the last two periods of recovery, this moves from the beginning in modest levels to consider it as an attested conjecture. On another side, though Random Forest evaluation goes in more acceptable graded layers, their values twist from one phase to the next consecutive. In the end, there is a difference of 6% between the recovery points at 3 and 12 months, having the latter with a higher accuracy, which hints at a degradation in the studied language.

Focusing on the annotations for the 'aux' and 'modals' components progress through the reviewed five recovery stages, notice that at 3, 12, and 24 months, Random Forest and SPAARC evaluations curves act as a mirror. Such fact is not replicated at 6 months recovery point where they behave comparatively as the opposite. Additionally, at the 9-month recovery stage, Random Forest evaluation coincides for both 'aux' and 'modals' whose value is near to the average resulting of the values obtained for the corresponding SPAARC evaluation, only 1% apart. Finally, except for the recovery stage at 24 months, the efficacy of both features fluctuates in a [83.11, 95.45] range.

To complement the analysis of the assessment step, the Bayes Net evaluation curve is added to the learning algorithms previously explained. There, a relatively uniform answer from the whole examined 7-feature set, both as a whole and individually, was observed. It is clear a drastic degradation until falling drastically at 9 months. The remaining responses move modesty in a [50.57, 77.27] interval, which, again, does not provide sufficient information to support any approximation from it.

2.5.4 RESULTS FOR SYNTAX PRODUCTIVITY

The shortened response of the last defined subset, $\{N_1, V_3, S_{17}\}$, corresponding to the learning algorithms evaluation to carry out an analysis of 'Productive Syntax' is also presented in graphs (Figures 2.4, 2.5, 2.6), where the evolution of the curves illustrates the efficacy to discriminate the studied group from the negative sample across the comprised period. They consist of the behavior curve for every single component as well as the response curve as a set for every stage of recovery. Attributed to the natural complementarity of the established subgroup, it was hypothesized that the 3-feature-set IPSYN evaluation curve will remain over any other individual evaluation curve. Figures 2.4, 2.5, 2.6, clarify that the trajectory of the 3-feature-set IPSYN reaction mirrors the N_1 behavior across the whole period, which at the same time stays above V_3 and S_{17}. On top of that, the three, learning algorithms showed similar behavior.

A line chart exhibiting a more stable reaction from 3 to 24 months at different stages of recovery after the injury. After evaluating attributes with the Random Forest algorithm, we have a few coincidences between V3 and the extension of it with characteristic S17. Besides, full replication of N1 into the response as a whole. The core for sentence structure, the S17 curve moving still in an acceptable scale below 90%, remains below the rest.

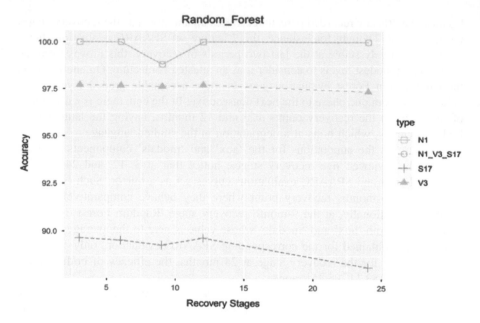

FIGURE 2.4   Random forest accuracy, seven IPSYN-feature set.

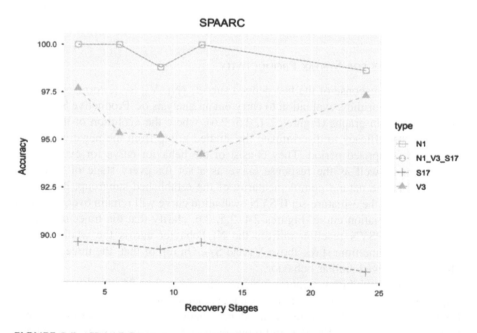

FIGURE 2.5   SPAARC accuracy, seven IPSYN-feature set.

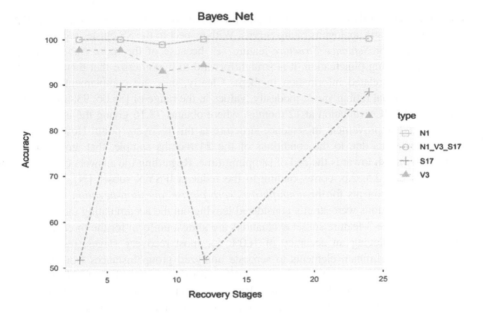

FIGURE 2.6 Bayes net accuracy, seven IPSYN-feature set.

A line chart exhibiting a quite consistent response for 'noun-phrase', 'verb-phrase', and 'sentence-structure' representative variables, 'N1', 'V3', and 'S17', in addition to the combination of them, '{N1, V3, S17}'. SPAARC algorithm generates a curve corresponding to the latter mirroring the 'N1' response. 'V3' curve shows a distinguished descension from three to twelve recovery stages to then turn up in ascension close to the initial accuracy assessment. 'S17' moves discretely below 90% accuracy levels but is stable.

A line chart exhibiting individual and group changes at periodical time points for 'N1', 'V3', 'S17' representing features evaluated with Bayes Net. 'S17' curve shows a similar response for 3, 6, and 24 months near 90% of efficacy, but a drastic decline at 12 months matching the initial value few above 50% accuracy. 'N1' and response as a whole mirror each other across the complete recovery stages. 'V3' curve variates in steady descending.

2.5.5 ANALYSIS AND DISCUSSION OF RESULTS FOR SYNTAX ANALYSIS

The first thing to remark is the fact that the group QN, representing the indices for *questions* and *negations* rules, is contributing minimal information, in general. These have levels of evaluations hardly surpassing 50% or even staying flat, i.e., the same for every recovery stage. On the opposite side is the NP collection, corresponding to the *Noun-Phrasal*-feature set whose accuracy is 100% at the 3-month recovery point and remains close to it for the rest of the period, indicating there is minimal alteration with respect to their production, i.e., noun-phrases, in the observed language. Regarding the VP set, which is the *verb phrase*, whose elements contribute among them, having an evaluation curve, as a set, staying predominantly on top of the single analogous rating curves. With respect to the SS cluster representing the measures for *Sentence Structure*-feature set, because of the more elaborated directives regulating punctuation, it is straightforward to hypothesize that there is more extensive inter-related information in them. Clearly, they are complementary since though individual ratings move modestly, values in the range of [85.06, 93.42], except for the SPAARC evaluation at 12 months, where obtains 78.16 giving the impression of a marked improvement in sentence structure at this recovery point. However, we must be cautious due to the conditions of the 24 months sample, that amelioration appears to vanish towards the end of sampling time. Regarding the answers condensed in Figures 2.4, 2.5, 2.6, corresponding to the reduced IPSYN subset $\{N_1, V_3, S_{17}\}$, where the components for the *noun-phrase, verb-phrase, question-negation, sentence-structure* collections were strictly considered leaving out the accumulating indices, it is observed that the 3-feature subset evaluations are almost uniform for the three learning algorithms, achieving an accuracy of 100% for most recovery stages, resulting in substantial discriminant elements to separate analyzed group instances from the negative samples. However, this efficacy nullifies an exhaustive analysis to identify the minimal amelioration or decline in the TBI-affected language responses. Furthermore, it seems that the overall behavior is dominated by N_1, the *noun-phrase* component. Proceeding with the examination of the separated reaction for S_{17} and V_3, a replication of a steady response from one between SPAARC and Random Forest is observed for the representative *sentence structure* group, indicating possible improvement in the altered language. This was in contrast with an unstable answer shown in the Bayes Net evaluation curve, which would be difficult to interpret.

About the V_3 assessment, Random Forest values show this component to be a valid judgment to discriminate positive from negative cases, whereas a flat curve might draw nothing about the inspected language progress or might describe there is not an improvement in syntax production with respect to the *verb phrase* set. While, for this element, SPAARC relatively coincides with Bayes Net assessment, except for the last recovery point, indicating a probable change for the better, in the examined language status.

2.6   F1-SCORE EVALUATION

The results presented above were summarized in terms of accuracy measure, next, they are detailed in terms of their corresponding F1-score. One indicator to select the learning algorithms was to find well-balanced study cases in comparison with the negative sample correctly tagged to make certain accuracy be a close representation of how data is behaving. In the case of the CNNLA feature set, Table 2.1 shows percentages of accuracy in contrast to their corresponding F1-score to display how close they are. The mean of the differences in absolute value is 0.65 for SPAARC and 0.26 for Random Forest. Such balance does not occur with Bayes Net which is evident in the considerable variation between accuracy and F1-score, having a mean difference of 21.10. Relatively the same situations occurred for the *modals* component. In the case of the IPSYN feature group, though it did not make sense to revise 100 accuracies, so, the values corresponding to the S_{17} component,

TABLE 2.1

Accuracy and F1-score for the 'aux' feature with learning algorithms

Time point (months)	SPAARC Accuracy	SPAARC F1-Score	Random Forest Accuracy	Random Forest F1-Score	Bayes Net Accuracy	Bayes Net F1-Score
3	88.636	89.362	88.636	88.889	51.136	67.669
6	91.954	92.308	88.506	88.636	50.575	67.175
9	94.118	94.253	91.764	91.765	0	0
12	84.091	83.721	85.227	85.057	51.136	67.669
24	85.714	84.058	90.909	90.141	55.844	0

The mean differences by learning algorithm in the order presented are: 0.648, 0.264, 21.102.

TABLE 2.2

Accuracy and F1-score for the 'S_{17}' feature with learning algorithms

Time Point (months)	SPAARC Accuracy	SPAARC F1-Score	Random Forest Accuracy	Random Forest F1-Score	Bayes Net Accuracy	Bayes Net F1-Score
3	89.655	90.909	89.655	90.909	51.724	68.189
6	89.535	90.722	89.535	90.722	89.535	90.725
9	89.286	90.322	89.286	90.322	89.286	90.322
12	89.655	90.909	89.655	90.909	51.724	68.189
24	88.158	88.312	88.158	88.312	88.156	88.312

The mean differences by learning algorithm in the order presented are: 0.977, 0.977, 7.058.

which appears to be more sensitive to the current evaluation, are included in Table 2.2 where accuracy and F1-score corroborate their proximity with a mean of the absolute values of the differences, 0.98 for SPAARC and Random Forest, and 7.06 for Bayes Net. With this, F1-score information provides support to the observations that were done in the discussion sections.

2.7 A SENSITIVITY STUDY

The next phase in our analyses is to establish a subset of characteristics maintaining strength to data variation but simultaneously showing sufficient sensitivity to reflect subtle alterations in the sample across the whole group of data points evaluated. After observing results from the $\{N_1, V_3, S_{17}\}$ feature subset, it was evident the exploration of how sensitive the binary composition $\{V_3, S_{17}\}$ could be. This is summed up in Figure 2.7 corresponding to the Random Forest algorithm and in Figure 2.8 for

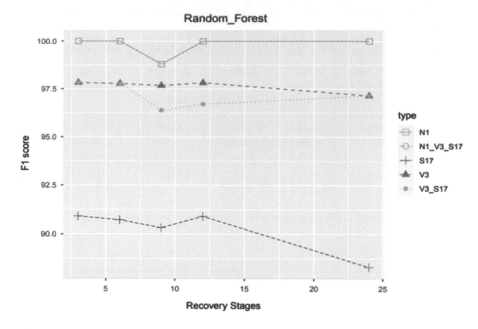

FIGURE 2.7   Random forest F1-score, seven IPSYN-feature set.

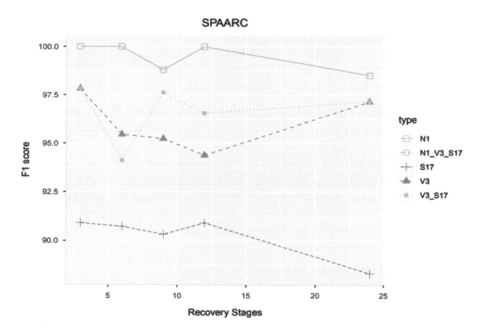

FIGURE 2.8   SPAARC F1-score, seven IPSYN-feature set.

SPAARC which encapsulate individual responses in addition to the two- and three-members subsets, showing this time F1-score to observe them comparatively. It appears that both learning algorithms reveal certain improvements at 12 months for the first recovery stage and that a part of it is preserved at the last time point monitored.

A line chart exhibiting fewer fluctuating changes from 3 to 24 months at different stages of recovery after the injury. Random Forest algorithm is the instrument to evaluate IPSYN features, 'N1', 'V3', and 'S17'. The values tabulated are F1-score. Finding some agreement between 'V3', which was combined with characteristic 'S17'. While a full reproduction of the responses as a whole and singly 'N1' contrasts the behavior of 'S17' curve which varies still in an acceptable scale around 90%, but remaining fall apart from the rest.

A line chart exhibiting a still unsteady response, though less sharpy for the variables, 'N1', 'V3', and 'S17' representing the 'noun-phrase', 'verb-phrase' and 'sentence-structure' clusters, in addition to, the reaction of the pair '{V3, S17}' and as a whole. With SPAARC algorithm, we have a curve of the evaluation as a whole, corresponding point to point to the 'N1' response. 'V3' curve descends from three to twelve time points and ascends close to the initial accuracy assessment at the last phase of recovery. The '{V3, S17}' set moves in opposite direction at each recovery stage resting at the same 'V3' index value at the last recovery point. 'S17' moves stably separately around 90% F1-score levels.

2.8   CONCLUSIONS

In this chapter, two sets of features to comprise both a lexical and a productive syntax were studied. Expanding the feature analysis based on a correlation approach, which could be considered as accepted from the linguistic point of view, together with a learning algorithms evaluation, shows evidence of augmenting the information that could be extracted from a given sample set, as it was noticeable with the process for the CNNLA feature set. This same extension applied to the defined rules used for assessing productive syntax, determining that *noun-phrase, verb-phrase,* and *sentence-structure* directives might be together, a significant instrument to discriminate TBI-language from negative samples.

However, a deep examination closer to an actual response in the modifications that inspected language suffers at each phase of recovery demands going into single-feature evaluation tier not before a substantial subset has been demarcated. A non-surprised finding is that NP, VP, SS directives are condensing the suitable information revealed, as they were linguistically defined (Scarborough 1990, MacWhinney 2000, 2020). It was also determined that *question-negation* rules, under the carried-out analysis, are not truly informative for any of the two purposes, i.e., discriminative or response analysis as reported here, a fact that could be attributable to the sort of tasks the analysis was based on.

Regarding the learning algorithms, SPAARC and Random Forest showed a relative consistency in their evaluations for both CNNLA and IPSYN feature sets, in contrast to Bayes Net, which is relatively more appropriate for a syntax analysis, where it shows a quite unresponsiveness to the variation of the data but not suitable for lexical analysis.

Based on the summarized findings, it can be concluded that language technologies and statistical analysis are boosting one another to set out an examination of the impaired language faculties, as is the case of the TBI-affected language.

BIBLIOGRAPHY

Coelho, C. A., B. Grela, M. Corso, A. Gamble, and R. Feinn. 2005. Microlinguistic deficits in the narrative discourse of adults with traumatic brain injury. In: *Brain Injury* 19, no. 13: 1139–1145.

Elbourn, E., B. Kenny, E. Power, C. Honan, S. McDonald, R. Tate, A. Holland, B. MacWhinney, and L. Togher. 2018. Discourse recovery after severe traumatic brain injury: Exploring the first year. In: *Brain Injury* 33, no. 2: 143–159. doi: 10.1080/02699052.2018.1539246

Fraser, K., J. A. Meltzer, N. L. Graham, C. Leonard, G. Hirst, S. E. Black, and E. Rochon. 2013. Automated classification of primary progressive aphasia subtypes from narrative speech transcripts. In: *Cortex* 55, pp. 43–60. doi: 10.1016/j.cortex.2012.12.006

Fraser, K. C., G. Hirst, N. L. Graham, J. A. Meltzer, S. Black, and E. Rochon. 2014a. Comparison of different feature sets for identification of variants in progressive aphasia. In: *Workshop on Computational Linguistics and Clinical Psychology: From Linguistic Signal to Clinical Reality*, pp. 17–26. Association for Computational Linguistics. doi: 10.1016/j.cortex.2012.12.006

Fraser, K. C., G. Hirst, J. A. Meltzer, J. E. Mack, and C. K. Thompson. 2014b. Using statistical parsing to detect agrammatic aphasia. In: *Proceedings of the 2014 Workshop on Biomedical Natural Language Processing (BioNLP 2014)*, pp. 134–142.

Gabani, K., M. Sherman, T. Solorio, Y. Liu, L. M. Bedore, and E. D. Peña. 2009. A Corpus-based approach for the prediction of language impairment in monolingual English and Spanish-English bilingual children. Human language technologies. In: *Proceedings of human language technologies: the 2009 annual conference of the North American chapter of the Association for Computational Linguistics* (pp. 46–55).

MacWhinney, B. 2000. *The CHILDES Project, Tools for Analyzing Talk*. 3rd ed. Mahwah, NJ.: Lawrence Erlbaum Associates.

MacWhinney, B. 2020. *Tools for Analyzing Talk – Electronic Edition Part 2: The CLAN Programs*. Pensilvania: Carnegie Mellon University. doi: 10.21415/T5G10R

Oliva, J., J. I. Serrano, M. D. del Castillo, and A. Iglesias. 2014. A methodology for the characterization and diagnosis of cognitive impairment – Application to specific language impairment. In: *Artificial Intelligence in Medicine* 61: 89–96.

Pakhomov, S. V. S, G. E. Smith, S. Marino, A. Birnbaum, N. Graff-Radford, R. Caselli, B. Boeve, and D. S. Knopman. 2010. A computerized technique to assess language use patterns in patients with frontotemporal dementia. In: *Neurolinguistics* March 1; 23, no. 2: 127–144. doi: 10.1016/j.jneuroling.2009.12.001

Peintner, B., W. Jarrold, D. Vergyri, C. Richey, M. A. Gorno-Tempini, and J. Ogar. 2008 Learning diagnostic models using speech and language measures. In: *30th Annual International IEEE EMBS Conference Vancouver*.

Power, E., S. Weir, J. Richardson, D. Fromm, M. Forbes, B. MacWhinney, and L. Togher. 2020. Patterns of narrative discourse in early recovery following severe traumatic brain Injury. In: *Brain Injury* 34, no. 1: 98–109.

Rentoumi, V., G. Paliouras, D. Arfani, K. Fragkopoulou, S. Varlokosta, E. Danasi, and S. Papadatos. 2017. Automatic detection of linguistic indicators as a means of early detection of Alzheimer's disease and of related dementias: A computational linguistics analysis. In: *CogInfoCom*, pp. 11–14.

Scarborough, H. S. 1990. Index of productive syntax. In: *Applied Psycholinguistics* 11, no. 1: 1–22.

Solorio, T. and Y. Liu. 2008. Using Language Models to Identify Language Impairment in Spanish-English Bilingual Children. In: *BioNLP 2008: Current Trends in Biomedical Natural Language Processing* pp. 116–117.

Steel, J. and L. Togher. 2018. Social communication assessment after TBI: A narrative review of innovations in pragmatic and discourse assessment methods. In: *Brain Injury* 33. no. 1: 48–60. doi: 10.1080/02699052.2018.1531304

Stubbs, E., L. Togher, B. Kenny, D. Fromm, M. B. F. Margaret, S. McDonald, R. Tate, L. Turkstra, and E. Power. 2018. Procedural discourse performance in adults with severe traumatic brain injury at 3 and 6 months post injury. In: *Brain Injury* 32, no. 2: 167–181. doi: 10.1080/02699052.2017.1291989

TalkBank. 2021. TalkBank. https://www.talkbank.org/

TBIBank. 2021. TBIBank. https://tbi.talkbank.org/. Accessed: March 3rd, 2021.

Thompson, C.K. 2013. Northwestern narrative language analysis (NNLA) theory and methodology. Evanston, IL: Aphasia and Neurolinguistics Research Laboratory Northwestern University.

Young, E. 2017. *Discourse Changes Following Severe Traumatic Brain Injury: A Longitudinal Study*. Electronic Theses and Dissertations Fall 11–13–2017. Available from: https://digitalrepository.unm.edu/shs_etds/19

3 Natural Language Processing for Mental Disorders: An Overview

Iacer Calixto
Amsterdam University Medical Centers, University of
Amsterdam, Amsterdam, The Netherlands

Victoria Yaneva
University of Wolverhampton, Wolverhampton,
United Kingdom

Raphael Moura Cardoso
University of Brasília, Brasília, Brazil

CONTENTS

DOI: 10.1201/9781003138013-3

3.1 INTRODUCTION

According to the World Health Organization (WHO 2018), mental health refers to "a state of well-being in which the individual realises his/her own abilities, can cope with the normal stresses of life, can work productively and fruitfully, and is able to make a contribution to the community". In other words, mental health does not just mean the absence of mental disorders. Moreover, the WHO estimates that one in four people using health services suffer from at least one psychiatric or behavioural disorder, and global costs related to mental health are expected to reach US$ 6 trillion by 2030, which is more than double compared to 2010 (Mnookin 2016).

The increasing economic and societal costs of mental disorders worldwide have motivated an increasing body of work on how to use different technologies to assist people living with these conditions. In this chapter, we provide an overview of the Natural Language Processing (NLP) applications and datasets dedicated to addressing problems related to mental health in the NLP literature. We focus on the different applications proposed, the types of data sources these applications use, and the languages they cover.

We first lay out a few crucial definitions we use throughout this chapter. *Psychiatric and behavioural disorders* refer to conditions that negatively affect an individual's mood, cognition or behaviour, and impair social life as well as daily activities. Specifically, *psychiatric disorders* refer to conditions or syndromes that are thought to be the result of some brain dysfunction (e.g., schizophrenia, depressive disorder) and *behavioural disorders* (also known as *psychological disorders*) refer to conditions whose symptoms are primarily observed at the behavioural level (e.g., anorexia, addiction). However, this distinction is blurred because there is no clear-cut evidence of the causal role of the brain or genetic factors for some disorders, which does not mean that such factors do not contribute to the disorder's manifestation (Holmqvist 2013; Borsboom et al. 2019). An example of the difficulty of distinguishing between the two is Autism Spectrum Disorder (ASD), which is considered a developmental disorder of neural origin (American Psychiatric Association 2013), but is still often classified as a behavioural disorder. Since the scientific debate around these definitions and classifications remains outside of the scope of this chapter, we use the term *mental disorders* to refer to any psychiatric or psychological disorders, including neurodevelopmental disorders, and in Section 2.1 we discuss the specific list of conditions we investigate under this umbrella term.

Mental disorders are diagnosed by a psychiatrist or psychotherapist based on a clinical assessment and the patient's self-report (or a third-party report) of specific symptoms and signs. Since this diagnosis is based on behavioural assessment, it is often subjective, may require long periods of observation, and is costly to obtain. These drawbacks have spurred a research line of identifying objective markers of these conditions in behavioural data (e.g., speech, text, or eye-tracking data) to provide additional, objective evidence to the diagnostic process. Another line of research associated with improving the quality of life of people who live with mental disorders is the development of assistive technology such as conversational agents to help people with depression, reading comprehension support tools for people with autism, and spell-checkers that are able to capture dyslexia-specific errors, among others. The development of such diagnostic and assistive tools is a vibrant research area in different technology-related fields, many of which include the processing of language and thus use approaches from the NLP field.

NLP is an inter-disciplinary field applying methodology of computer science and linguistics to the processing of natural languages (Jurafsky and Manning 2012). NLP methods are applicable to clinical assessments written by psychiatrists and psychotherapists (e.g., in electronic health records; EHRs), as well as to utterances or texts produced by patients. Even when not in a clinical or therapeutic setting, e.g., when one writes posts on social media, this information could still be valuable and used to detect *linguistic markers* associated with specific mental disorders, as well as to drive the development of assistive technology.

3.1.1 Main Contributions

Our main contributions in this work are: i) We provide a birds-eye view on the current state of the research being conducted by the NLP research community on topics related to mental disorders with a focus on non-English and low-resource languages; ii) We survey a detailed list of the resources and datasets available for NLP researchers to train and validate data-driven models that address tasks related to mental disorders; and iii) Given the current landscape, we wrap up with suggestions and recommendations on the most important next steps to accelerate and fully realise the potential NLP brings to mental health with special attention to ethical and privacy issues.

3.1.2 Related Surveys

We note that there are a number of related surveys relevant to our topic in this Chapter: Névéol et al. (2018) provide an overview of clinical NLP for languages other than English, where the focus is on *clinical data*, e.g., EHRs; Graham et al. (2019) survey applications of artificial intelligence to mental health, which includes not only NLP but also other fields (e.g., using computer vision to predict mental disorders from brain images); Bzdok and Meyer-Lindenberg (2018) provide an overview on machine learning applied to psychiatric disorders, covering broad *biological markers* (e.g., genetic, behavioural, neural) but not linguistic markers. Finally, Šuster et al. (2017) discuss issues in clinical NLP, such as the difficulty to access high-quality data due to privacy concerns, and highlight different possible sources of bias in data. Our survey differs from previous work in that it draws a landscape of how mental disorders are investigated within the NLP community in particular, with a focus on the different applications, data sources, and languages covered by the NLP literature on mental disorders. Our target audience is NLP researchers and professionals interested in methods, applications, resources and datasets broadly related to mental disorders.

3.2   THE LANDSCAPE OF NLP FOR MENTAL DISORDERS

Our goal in this Section is to provide a high-level, *birds-eye overview* of the area or, in other words, a glimpse into the overall landscape of the research in NLP for mental disorders. To do that, we survey papers related to this topic using the criteria described below. We implement this procedure in software and make our code publicly available[1]. Our search includes papers indexed by the *Association for Computational Linguistics (ACL) Anthology* (Gildea et al. 2018)[2], arguably the database with the most important scientific venues in NLP. This way we provide an overview of how research on mental disorders is approached in top-tier NLP journals, conferences, as well as dedicated workshops.

3.2.1   INCLUSION AND EXCLUSION CRITERIA

We include publications that conform to the following inclusion criteria: 1) Publications must be indexed by the ACL Anthology, i.e., an indexed journal, conference or workshop, without restrictions on the publication year; 2) Publications must include in their title or abstract at least one of the keywords in our keywords list, which we discuss next. Criteria (1) and (2) above are automatic and implemented in software, i.e., one can directly download our code and run it to reproduce the output of criteria (1–2) using the same keywords we use. We now discuss each criterion in more detail.

3.2.1.1   ACL Anthology (criterion 1)

We use an authoritative index to search for the NLP publications to include in this survey. As stated on the ACL public website, the Association for Computational Linguistics "is the premier international scientific and professional society for people working on computational problems involving human language, a field often referred to as either computational linguistics or natural language processing (NLP)"[3]. The ACL Anthology indexes 62, 486 papers across journals, conferences, workshops, and shared tasks, and is arguably the most important index for research publications in NLP.

3.2.1.2   Keyword Matching (criterion 2)

We use the exact substring match between a keyword and the title and abstract, and a paper is considered a match if a keyword is contained in either of the two. We choose keywords to identify papers dealing with mental disorders that have a high estimated burden and are the most epidemiologically prevalent (Whiteford et al. 2015). We use the keywords: "autism", "ASD"[4], "dyslexia", "depression", "obsessive compulsive disorder", "obsessive-compulsive disorder", "OCD"[5], "attention deficit and hyperactivity disorder", "ADHD"[6], "schizophrenia", "down syndrome", "substance abuse", "tobacco abuse", "tabagism", "eating disorder", "anorexia", "bulimia", "mental health" and "developmental disorder".

3.2.1.3   Manual Validation

Next, we manually validate all papers by reading their titles and abstracts and filtering out any incorrectly retrieved papers (false positives), leaving a total of

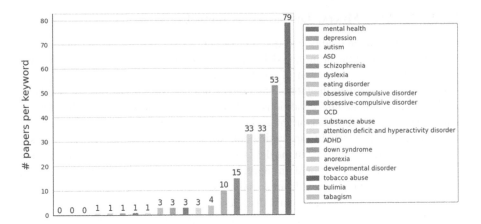

FIGURE 3.1 Number of retrieved papers from the ACL anthology with each keyword. Keywords "tabagism", "bulimia", and "tobacco abuse" retrieved zero papers.

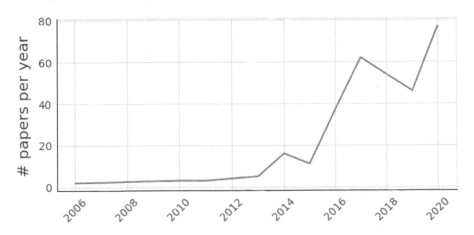

FIGURE 3.2 Number of retrieved papers per year (2006–2020).

155 unique publications. In Figure 3.1, we show the number of papers retrieved per keyword and note that the counts shown can include the same paper retrieved with different keywords. For example, there is a large overlap between papers retrieved with keywords "autism" and "ASD", as expected. Overall, the keyword which matches the largest number of papers is "mental health" (79), which in practice is used as an umbrella term for mental health-related work broadly. The keywords "depression", "autism", "ASD" and "schizophrenia" come next, matching a total of 53, 33, 33 and 15 papers, respectively. Conversely, some keywords matched zero publications: "bulimia", "tobacco abuse" and "tabagism".

In Figure 3.2, we show the number of papers retrieved per year of publication. We note that there is a clear surge in the number of publications on mental health-related topics in the last five years in NLP venues.

3.2.2   NLP for Mental Disorders: A Birds-Eye View

The trade-off between quality and quantity of data exists in many fields and domains but is especially prominent in the NLP literature related to mental disorders. This can be explained by a combination of multiple factors, such as the difficulty to access participants with verified diagnoses, challenges in data collection further exacerbated by the specifics of certain conditions and, importantly, the ethical implications of making such data publicly available. These restrictions have resulted in some NLP studies using data collected from participants with verified mental disorder diagnoses, where there is high confidence in the data sources but a small amount of data available. One way in which the NLP field has approached this data scarcity problem is through obtaining large amounts of text from social media such as Twitter, Reddit, and others, by scraping content related to mental disorders. In these cases, the abundance of data allows for the application of state-of-the-art NLP techniques but this advantage comes at the high cost of not knowing whether the data comes from participants who indeed have a certain condition, even in cases where the participants themselves think that they do. This foundational distinction between the data sources (verified vs. unverified diagnoses) entails different dataset sizes, which in turn results in different methods, applications, and levels of confidence in the results.

One of the first findings of our review is that the use of these two types of datasets (verified vs. unverified diagnoses) vary across conditions. For example, more narrowly specified conditions such as *autism, dyslexia* or *schizophrenia* are associated with more papers that use data from verified participants, while broader umbrella terms such as *mental health* or *depression* retrieve more papers using social media data. This means that the knowledge gained by many NLP studies in mental health comes from unverified sources, which is a tendency worth noting when discussing the implications of this research.

Below, we outline studies on different mental disorders found in the NLP literature, where we specifically highlight: i) the types of applications, ii) their data sources, and iii) language coverage. In some cases, papers returned under one category (e.g., *mental health* or *developmental disorders*) may, in fact, refer to a more specific category (e.g., *depression* or *autism*), in which case they are discussed under their more specific heading. The same goes for papers that introduce a resource, which may be returned under *mental health* for example, but are discussed in a specific section dedicated to resources (Section 2.3).

3.2.2.1   Autism Spectrum Disorder (ASD)

The majority of papers on ASD use data from verified participants, but datasets are seldomly made publicly available, meaning that there are no replications or reproductions of any of these studies. The main application of interest is the detection of autism at an early age owing to the difficulties of diagnosing the condition (e.g., Tanaka et al. 2014; Parish-Morris et al. 2018; Hauser et al. 2019; Heeman et al. 2010). The majority of these papers focus on the identification of linguistic markers in transcribed speech, often recorded during the diagnostic process, and often in the form of a dialogue between the person with autism and an interlocutor. Other studies that use data from participants with confirmed diagnoses focus on readability research (Yaneva et al. 2017) and complex word identification (Štajner et al. 2017), including a survey on the text adaptation needs of adults with ASD (Yaneva et al. 2019) as well as an analysis of visual-semantic priming and narratives (Regneri et al. 2020). Yaneva et al. (2016) publicly releases a multimodal dataset with text and eye-tracking data collected from adults with and without autism. Finally, among the papers that use data from participants with unverified diagnoses, one study focuses on topic modelling for an ASD support forum (Ji et al. 2014), while other studies focus on text simplification or reading assistance without reporting evaluation with ASD participants (Barbu et al. 2013; Evans et al. 2014). Among all studies on autism, only Tanaka et al. (2014) explores a language that is not English (i.e., Japanese). This results in a serious gap in our knowledge of how linguistic markers of autism are expressed in different languages, whether interlinguistic commonalities exist between these markers, and whether there are common ways to overcome the challenges people with autism may face when reading in different languages.

3.2.2.2 Dyslexia

Unlike the autism studies above, the papers focused on dyslexia explore a greater variety of languages, including Spanish, French, German, Arabic, and Icelandic. Our search for dyslexia returned ten studies, out of which six use data from participants with verified diagnoses. These include three resource papers with annotated writing errors made by dyslexic participants in Spanish (Rello et al. 2014a), German (Rauschenberger et al. 2016) and Arabic (Alamri and Teahan 2017). Other studies focus on the evaluation of reading support tools for dyslexia such as testing whether highlighting the main ideas or including graphical schemes has an impact on readability and comprehensibility for Spanish (Rello et al. 2012, 2014b), or whether lexical simplification helps dyslexic children read French (Gala and Ziegler 2016). The studies that do not rely on participants with confirmed dyslexia are related to automatically creating a set of errors for Icelandic to train a spell checker that is hypothesised to help people with dyslexia (Friðriksdóttir and Ingason 2020), the automatic detection of letter combinations in Shakespeare sonnets which may be found challenging by people with dyslexia (Bleiweiss 2020), and developing a reading tutor for children with dyslexia that is yet to be evaluated (Ward and Crowley 2011). Quiniou and Daille (2018) propose a method to detect constructions in French annotated by a speech therapist that may be challenging for people with dyslexia.

3.2.2.3 Schizophrenia

Similarly to autism, the majority of papers using data from participants with a verified diagnosis of schizophrenia focus on identifying linguistic markers in speech samples that are then used to detect the condition (e.g., Gutiérrez et al. 2017; Just et al. 2019). This is done predominantly for English, with some exceptions being Bulgarian (Boytcheva et al. 2017), French (Amblard et al. 2020), and Hebrew (Bar et al. 2019). Another line of work proposes extracting symptoms from EHRs that could indicate the onset of the condition (Viani et al. 2018; Gorrell et al. 2013). One paper aims to predict the probability of adherence to assigned treatment using dialogues from outpatient consultations (Howes et al. 2012). Finally, several papers mine social media such as Twitter for linguistic markers of schizophrenia based on self-reported diagnosis (e.g., Mitchell et al. 2015; Loveys et al. 2017).

3.2.2.4   Depression

The papers related to depression are the second largest group returned by our search. Among them, 35 studies use data from participants with unverified diagnoses (predominantly from Twitter or Reddit), and 8 use data from participants with verified diagnoses. In the majority of cases, the papers' main goal is to detect the depressive condition (23 papers from the unverified group and six papers with verified participants). This is done through speech analysis (e.g., Lamers et al. 2014) in the cases of papers with verified participants and social media text in the case of papers without verified participants (e.g., Mowery et al. 2016; Song et al. 2018; Husseini Orabi et al. 2018; An et al. 2020). Some studies use both types of data sources, for instances Delahunty et al. (2019). In terms of methods, most papers directly model the condition by approaching its detection as a classification task, with a few offering more knowledge-rich approaches such as semi-automatic emotion annotation (Canales et al. 2016), detecting the use of figurative language in text as an auxiliary task (Yadav et al. 2020), and topic modelling for depression detection (Resnik et al. 2015).

Besides depression detection, other applications include the investigation of patients' attitudes to treatment (Mikal et al. 2017), discourse analysis for support groups (Yip 2018), and quantifying the risk of self-harm (Yates et al. 2017). We note that for some of these applications, e.g., discourse analysis for support groups, the lack of a verified diagnosis is not necessarily a limitation. These studies make a valuable contribution to our understanding of the concerns and coping strategies of people who feel depressed, which can be used as a source of great value in therapy or other social settings. Nevertheless, the lack of a verified diagnosis remains an issue when trying to detect the condition. This is further exacerbated by the fact that many of these studies use a very broad, non-scientific definition of depression, which does not allow verifying the actual usefulness of the tool and its ability to discriminate depression from other conditions, which may have similar linguistic markers (e.g., anxiety, bipolar disorder). Another serious limitation of NLP research related to depression is the lack of linguistic diversity. Studies such as Santos et al. (2020) who study Brazilian Portuguese are a rare exception in this English-dominated sub-domain.

3.2.2.5   General Mental Health

The papers returned for the *mental health* query are the largest category of papers and are fairly homogeneous in terms of data sources in languages. The vast majority use English-language social media data (e.g., Jamil et al. 2017; Mikal et al. 2017; Ireland and Iserman 2018; Loveys et al. 2018) to detect generic mental health concerns, depression or anxiety (often without following a formal definition of these conditions). Some studies such as Gkotsis et al. (2016a) focus on detecting suicide ideation in social media. The few exceptions using data from verified participants include the prediction of adult mental health from childhood writing (Radford et al. 2018), which was part of the 2018 CLPsych shared task, several papers mining electronic health records for information such as suicidality (Gkotsis et al. 2016b) and medication (Chaturvedi et al. 2020). Lamers et al. (2014) are an exception where speech features are used to predict depression using data from therapeutic interventions of (verified) patients living in a nursing home in Amsterdam (i.e., Dutch language). Regarding papers on languages other than English but without verified participants, one uses social media data to predict temporal mental health dynamics (Tabak and Purver 2020), and two propose conversational agents: one for Arabic to be used to help clinicians in administering treatment (Fadhil and AbuRa'ed 2019), and one for Japanese where the goal is to engage senior people using attentive listening techniques (Lala et al. 2017). While the body of work that uses social media to detect mental disorders has obvious limitations entailed by the lack of verified diagnoses, these studies are valuable for identifying new linguistic markers that can, at a later point, be evaluated using controlled datasets. Finally, Amir et al. (2019) use social media data to identify mental health issues "at large" and compare it with prevalence reports.

The applications developed for each of these conditions depend on and are driven by the available resources. To better characterise the relationship between the two, in the next section we present the datasets publicly available in the field of NLP for mental disorders.

3.2.3 DATASETS AND RESOURCES

This section describes datasets related to mental disorders that are available to the research community. The criteria for including these resources is that the papers describing them specifically state that they are available to other researchers, either through a repository or by contacting the authors and signing a data usage agreement. We do not include papers describing and using resources that were not noted as publicly available.

As can be seen in Tables 3.1 and 3.2, the number of resources that use data from participants with verified diagnoses and are publicly available is much smaller than those that use data from social media. Moreover, datasets from participants with verified diagnoses are typically very limited in size (as expected), such as the dataset released by Yaneva et al. (2017) with 27 texts, or Rauschenberger et al. (2016) with 47 texts. Albeit smaller, the resources using data from verified participants are comparatively more diverse in terms of the languages (English, Spanish, German) and conditions (autism, dyslexia, general mental health) they cover; by contrast, the social media resources that are publicly available typically target generic mental health related to anxiety, depression or suicide ideation and focus almost predominantly on English-language sources, notable exceptions being Park et al. (2020) for Korean and Lee et al. (2020) for Cantonese.

Next, in Section 3, among other things, we discuss the implications the availability of these resources have for the development of the field.

3.3 DISCUSSION

We draw three main conclusions based on the review of the papers presented in the previous sections. The first conclusion refers to the fact that the majority of papers in the field, especially the ones which refer to broad terms such as *mental health* and

TABLE 3.1

Summary of resources: Participants whose diagnosis was verified

Paper	Condition	Language	Description	Data Availability
Yaneva et al. (2016)	Autism	English	A multimodal data set of text and eye-tracking was data obtained from autistic and control readers. The data were elicited through reading comprehension tests and includes participants' responses to the questions. An extended version of the data set with 20 text passages and 60 questions is available in Yaneva (2016).	https://github.com/victoria-ianeva/ASD-eye-tracking-reading-corp
Yaneva et al. (2017)	Autism	English	A collection of 27 individual documents whose readability was evaluated by 27 different people diagnosed with autism	https://github.com/victoria-ianeva/ASD-Comprehension-Corpus
Rello et al. (2014a)	Dyslexia	Spanish	54 school essays and homework exercises provided by teachers from children and teenagers with dyslexia between 6 and 15 years old and 29 texts provided by parents with children with dyslexia with 887 misspellt words.	www.luzrello.com/dyslist.html (link not functional, January 22, 2021).
Rauschenberger et al. (2016)	Dyslexia	German	47 texts (homework exercises, dictations, and school essays) written by students from 8 to 17 years old with 1,021 unique errors	http://goo.gl/LRaUDA (link not functional, January 22, 2021).
Lynn et al. (2018)	General mental health	English	Data from the British Birth Cohort Study, which follows a cohort of children born in a single week in March 1958 in Great Britain. The data contains essays written by the children at age 11 and used to predict aspects of their mental health at various ages.	Data was made available as part of the CLPsych 2018 Shared Task.
Gratch et al. (2014)	Anxiety, depression and PTSD	English	Clinical diagnostic interviews conducted by humans, human-controlled agents, and autonomous agents. The participants include both distressed and non-distressed individuals. Data collected include audio and video recordings and extensive questionnaire responses; parts of the corpus have been transcribed and annotated for a variety of verbal and non-verbal features.	"Currently, the corpus is being shared on a case-by-case basis by request and for research purposes".

TABLE 3.2

Summary of resources: Participants with no verified diagnosis

Paper	Condition	Language	Description	Data Availability
Cohan et al. (2018)	General mental health	English	Posts extracted from a publicly available Reddit corpus corresponding to users who claimed to have been diagnosed with a mental disorder (20,406 users) and control users who are unlikely to have one of the mental disorders studied (335,952 users).	Available through a data usage agreement protecting the users' privacy.
Coppersmith et al. (2014)	General mental health	English	Tweets from people who have stated that they have depression (1 million), bipolar disorder (992k), PTSD (573k), seasonal affective disorder (421k), as well as a control group tweets (13.7 million).	Data was made available as part of the CLPsych 2015 Shared Task.
Park et al. (2020)	Suicide risk	Korean	2,791 posts from Korean Q&A site written by active-duty military personnel with 13,955 expert annotations of suicidal risk levels.	Available subject to research ethics agreement.
Anani et al. (2019)	General mental health	English	Shared task description paper that describes the RDoC (Research Domain Criteria framework) task in the BioNLP Open Shared Tasks. The dataset consists of Pubmed abstracts from papers related to mental health.	https://montana.app.box.com/s/kh0hmyn1jcj5ajvr2nibq4iwwgiv3 lefolder73225997826 invalid link May 7, 2021.
Matero and Schwartz (2020)	General mental health	English	A dataset collected from 1,900 Twitter users and annotated for affective language use with weekly and daily scores for six emotions.	https://github.com/MatthewMatero/AffectiveLanguageForecasting

(Continued)

TABLE 3.2 (Continued)
Summary of resources: Participants with no verified diagnosis

Paper	Condition	Language	Description	Data Availability
Yates et al. (2017)	Depression	English	A Reddit dataset consisting of users with self-reported depression diagnoses matched with control users.	"Available to researchers who agree to follow ethical guidelines".
Ghosh et al. (2020)	Suicide	English	Corpus of emotion-annotated suicide notes. The corpus consists of 2,393 sentences from around 205 suicide notes collected from various sources.	https://www.iitp.ac.in/~ai-nlp-ml/resources.html# CEASE
Lee et al. (2020)	General mental health	Cantonese	A corpus designed for training a virtual counsellor, which consists of a domain-independent sub-corpus that supports small talk for rapport building with users, and a domain-specific sub-corpus that provides material for a particular area of counselling.	Available for Research purposes on request to the first author.
Ellendorff et al. (2016)	General mental health	English	175 abstracts of PubMed research articles chosen based on a selection of mental disorders and annotated for events.	http://www.ontogene.org/current-pr/psymine
Milne et al. (2016)	General mental health	English	The ReachOut dataset consists of 65,024 youth-support-forum posts written between July 2012 and June 2015. The posts were annotated in four categories corresponding to levels of distress.	Data was made available as part of the CLPsych 2016 Shared Task.
Zirikly et al. (2019) and Shing et al. (2018)	General mental health	English	1,556,194 Reddit posts from 11,129 users collected from the r/SuicideWatch subreddit and annotated into four categories of suicide risk.	Data was made available as part of the CLPsych 2019 Shared Task.

depression, use data obtained from social media. Second, we note that there is a variety of NLP applications applied to mental disorders, starting with detection of various conditions based on linguistic markers, through the development of assistive technology of various kinds (e.g., conversational agents, reading tutors, text simplification), to topic modelling in support forums and social media. Finally, and somewhat unsurprisingly, the vast majority of the research in the field is done for the English language with very few exceptions focusing on other languages, and almost no studies on under-resourced languages. The paragraphs below discuss each of these findings and their implications for validity, ethics, and equity within the field of NLP for mental disorders.

3.3.1 Types of Data Sources

The fact that the majority of papers use social media data indicates that most of the knowledge in the field is derived from a population sub-sample who uses social media and does not have a verified diagnosis, meaning that there are serious issues related to selection bias and low outcome validity. While there are, undoubtedly, advantages to this approach given the amount of available data, there are also ethical considerations related to informed consent by vulnerable participants. The fact that this material is often public does not exempt researchers from the ethical standards that should guide the research and must be evaluated with extreme caution. For example, there is a potential privacy risk for the non-informed subjects: although data sanitisation is a minimum necessary technique for protecting an individual's privacy, this procedure may not always be sufficient to guarantee anonymity (Malin et al. 2013). To address the potential ethical issues stemming from the use of social media data from vulnerable populations (as is the case of those with mental disorders), we recommend that researchers not only place utmost importance on data anonymisation, but also only share the data subject to research ethics agreement whenever participant-produced language is used.

Having stated the drawbacks and risks of using social media data, we note that the collection of data from participants with verified diagnoses does not come without its own limitations, in addition to its scarcity. Such data can also have biases: clinical texts produced by the clinician and the patient can suffer from reporting bias; observational bias can also be an issue since not all variables involved in predicting a certain clinical outcome are written down in clinical texts; sampling bias can be a serious issue if certain subgroups of a population have limited access to medical or psychological services, etc. There is, therefore, no single best solution and awareness of the limitations of various data sources is crucial. Based on the reviewed literature, such awareness is not always demonstrated and these limitations are not always explicitly discussed.

3.3.2 Applications

While the applications developed within the field are valuable and diverse, a noticeable gap in many of these papers is the lack of discussion on how the research can be translated into public health policies and what risks and ethical implications this process would involve. As NLP technologies for mental disorders demonstrate great potential, we argue that NLP research on this topic should be better aligned to ethical principles guiding the application of AI technology in general (see Mittelstadt (2019)), which includes awareness of how the research will be used in practice and the mitigation of potential risks to ethical principles.

3.3.3 LANGUAGE COVERAGE

As noted above, the vast majority of papers returned by our search focused on the English language. This finding, especially since social media data is readily available in various languages, constitutes a serious gap within the field (i.e., see column "language" in Tables 3.1 and 3.2). This gap points to the existence of linguistic inequality in the development of NLP approaches to mental disorders, which could be somewhat rectified (at the very least) by the organisation of shared tasks using non-English social media data. This linguistic coverage gap in the publicly available datasets together with the known difficulties those with a linguistic and/or cultural minority background encounter to access mental healthcare (Kisely 2020; Mianji et al. 2020) make it clear that linguistic inequalities in mental healthcare are a major source of concern.

Another way to address the lack of NLP tools and resources related to mental disorders for non-English languages is to borrow techniques from other NLP domains. For example, a standard approach in NLP when working with a low-resource language is to leverage a machine translation system to translate the task resources available in English into the low-resource language or vice-versa (assuming a publicly available machine translation system for that language pair exists, e.g., Google Translate). This approach is popular in general NLP tasks, e.g., dependency parsing (McDonald et al. 2011) and named entity recognition (Mayhew et al. 2017). In clinical NLP, applications are varied and range from translating search terms for systematic reviews (Balk et al. 2013) to translating EHRs to aid patients who speak foreign languages in the United States (Liu and Cai 2015). For more examples of using machine translation in a clinical NLP setting, we refer the reader to the discussion in Névéol et al. (2018, p. 8).

Recent advances in natural language understanding suggest that a potentially better approach than applying machine translation is possible when dealing with low-resource languages: *cross-lingual transfer learning*. There are many publicly available pretrained multilingual language models (LMs), such as multilingual BERT (Devlin et al. 2019) and XLM-R (Conneau et al. 2020). These multilingual LMs usually cover about a hundred diverse languages (i.e., languages from many language families and with different scripts) and can be adapted relatively easily to different languages and/or tasks. The main idea is to continue training the multilingual LM using task data either in the low-resource language or in English, depending on what is available for the task at hand. Even when task data is available only in English, recent work has shown that pretrained LMs such as XLM-R work well in a *zero-shot cross-lingual* setting on many difficult NLP tasks, i.e., when training the model in one language (e.g., English) and performing inference with the model directly in another language (e.g., the low-resource language). State-of-the-art cross-lingual transfer learning approaches for traditional NLP tasks include performing an add-on training on an intermediate task in English (usually with a lot of data) before fine-tuning the model on the actual task of interest (Phang et al. 2020; Pruksachatkun et al. 2020), and using *adapters*, i.e., small sets of parameters that allow the model to be applied to different languages and tasks (Pfeiffer et al. 2020).

3.3.4 LIMITATIONS

One of the main limitations in this work has to do with our choices regarding which papers to include in our survey and which not to. We choose to only use the ACL Anthology as the database where to search for papers, both as a proxy to the "NLP research community" and to keep the number of papers surveyed reasonable. This means we could have missed papers published by NLP researchers, i.e., part of the "NLP community", in venues indexed by other scientific communities. In future work, we plan to extend this search approach to other relevant databases such as the Association Computing Machinery (ACM) or PubMed, which would require the applications of additional filters to ensure that the selected studies use NLP.

We highlight that our intent is not to provide a systematic review. We choose to prioritise breadth of knowledge over depth, and therefore do not aim at discussing specific methods in detail but to provide a snapshot of the current research in NLP for mental health. A detailed look into the specifics of individual disorders is the subject of future work. We also note that the reasons we choose to include some mental disorders and not others are varied: we try to prioritise mental disorders with high burden and epidemiological prevalence (Whiteford et al. 2015), to include sub-areas under the authors' expertise, all while making sure we conform to space constraints. While the issues discussed in this paper reflect the background of the authors (i.e., psychology, NLP, and accessibility), it is possible that some relevant perspectives of professionals from other fields have been omitted or underrepresented.

Overall, the reviewed literature presents an exciting picture of a young and promising multi-disciplinary field. Like any other such field, research on NLP for mental health grapples with problems related to resources, multilingualism, and the placement of its application into a complex real-world context. To solve these issues, the field can benefit from a closer alignment with established data collection and ethical principles in clinical psychology and the broader field of artificial intelligence, as well as a deeper exploration into the use of transfer learning for solving the existing linguistic inequalities within this domain.

3.4 CONCLUSIONS

We provided an overview of the current landscape of the research on the broad theme of NLP for mental health. To do that, we use the ACL Anthology, which is an authoritative database for scientific work in NLP. We survey papers that propose applications and datasets available to address problems related to mental health. The list of datasets provided is flagged as publicly available by their authors, and we note that even if the link to the data is no longer active, as in some cases, the data should in principle still be accessible upon contacting the original authors.

Our focus, in tandem with the core idea in this book, is to highlight solutions available in non-English and low-resource languages. We find that whereas the majority of papers only support English as we initially suspected, possible solutions to address this linguistic inequality include devising shared tasks on mental health-related topics using non-English languages, and more deeply investigating the application of state-of-the-art techniques from NLP such as cross-lingual transfer learning.

NLP technologies create many interesting possibilities to improve mental health services and diagnostics, but there is still a long way to go until this potential can be fully realised. As we seek to achieve a better alignment between the different disciplines that contribute to this type of research, NLP researchers should be mindful of the ethical concerns and linguistic inequalities within the current paradigm of this relatively young research area.

ACKNOWLEDGMENT

Iacer Calixto has received funding from the European Union's Horizon 2020 research and innovation programme under the Marie Skłodowska-Curie grant agreement #838188. Raphael Cardoso has received funding from FAPEG (#202010267000378). We thank our reviewers for their valuable comments.

NOTES

1 https://github.com/iacercalixto/nlp4health
2 For a full list of venues, see https://www.aclweb.org/anthology/
3 https://www.aclweb.org/portal/what-is-cl. Last access on February 1st, 2021.
4 Abbreviation for Autism Spectrum Disorder.
5 Abbreviation for Obsessive Compulsive Disorder.
6 Abbreviation for Attention Deficit Hyperactivity.

REFERENCES

Alamri, M., and W. J. Teahan. 2017. A new error annotation for dyslexic texts in Arabic. In *Proceedings of the Third Arabic Natural Language Processing Workshop*, pages 72–78, Valencia, Spain: Association for Computational Linguistics.

Amblard, M., C. Braud, C. Li, C. Demily, N. Franck, and M. Musiol. 2020. Investigation par méthodes d'apprentissage des spécificités langagières propres aux personnes avec schizophrénie (investigating learning methods applied to language specificity of persons with schizophrenia). In *Actes de la 6e conférence conjointe Journées d'Études sur la Parole (JEP, 33e édition), Traitement Automatique des Langues Naturelles (TALN, 27e édition), Rencontre des Étudiants Chercheurs en Informatique pour le Traitement Automatique des Langues (RÉCITAL, 22e édition). Volume 2: Traitement Automatique des Langues Naturelles*, pages 12–26, Nancy, France: ATALA et AFCP.

American Psychiatric Association. 2013. *Diagnostic and Statistical Manual of Mental Disorders* (5th ed.). Arlington, VA.

Amir, S., M. Dredze, and J. W. Ayers. 2019. Mental health surveillance over social media with digital cohorts. In *Proceedings of the Sixth Workshop on Computational Linguistics and Clinical Psychology*, pages 114–120, Minneapolis, Minnesota: Association for Computational Linguistics.

An, M., J. Wang, S. Li, and G. Zhou. 2020. Multimodal topic-enriched auxiliary learning for depression detection. In *Proceedings of the 28th International Conference on Computational Linguistics*, pages 1078–1089, Barcelona, Spain (Online): International Committee on Computational Linguistics.

Anani, M., N. Kazi, M. Kuntz, and I. Kahanda. 2019. RDoC task at BioNLP-OST 2019. In *Proceedings of The 5th Workshop on BioNLP Open Shared Tasks*, pages 216–226, Hong Kong, China: Association for Computational Linguistics.

Balk, E. M., M. Chung, M. L. Chen, L. K. W. Chang, and T. A. Trikalinos. 2013. Data extraction from machine-translated versus original language randomized trial reports: A comparative study. *Systematic Reviews*, 2, no. 1: 97.

Bar, K., V. Zilberstein, I. Ziv, H. Baram, N. Dershowitz, S. Itzikowitz, and E. Vadim Harel. 2019. Semantic characteristics of schizophrenic speech. In *Proceedings of the Sixth Workshop on Computational Linguistics and Clinical Psychology*, pages 84–93, Minneapolis, Minnesota: Association for Computational Linguistics.

Barbu, E., M. T. Martín-Valdivia, and L. A. Ureña-López. 2013. Open book: A tool for helping ASD users' semantic comprehension. In *Proceedings of the Workshop on Natural Language Processing for Improving Textual Accessibility*, pages 11–19, Atlanta, Georgia: Association for Computational Linguistics.

Bleiweiss, A. 2020. Neural transduction of letter position dyslexia using an anagram matrix representation. In *Proceedings of the 19th SIGBioMed Workshop on Biomedical Language Processing*, pages 150–155, Online. Association for Computational Linguistics.

Borsboom, D., A. O. Cramer, and A. Kalis. 2019. Brain disorders? not really: Why network structures block reductionism in psychopathology research. *Behavioral and Brain Sciences*, 42: e2.

Boytcheva, S., I. Nikolova, G. Angelova, and Z. Angelov. 2017. Identification of risk factors in clinical texts through association rules. In *Proceedings of the Biomedical NLP Workshop associated with RANLP 2017*, pages 64–72, Varna, Bulgaria: INCOMA Ltd.

Bzdok, D., and A. Meyer-Lindenberg. 2018. Machine learning for precision psychiatry: Opportunities and challenges. *Biological Psychiatry: Cognitive Neuroscience and Neuroimaging*, 3, no. 3: 223–230.

Canales, L., C. Strapparava, E. Boldrini, and P. Martínez-Barco. 2016. Innovative semi- automatic methodology to annotate emotional corpora. In *Proceedings of the Workshop on Computational Modeling of People's Opinions, Personality, and Emotions in Social Media (PEOPLES)*, pages 91–100, Osaka, Japan: The COLING 2016 Organizing Committee.

Chaturvedi, J., N. Viani, J. Sanyal, C. Tytherleigh, I. Hasan, K. Baird, S. Velupillai, R. Stewart, and A. Roberts. 2020. Development of a corpus annotated with medications and their attributes in psychiatric health records. In *Proceedings of the 12th Language Resources and Evaluation Conference*, pages 2009–2016, Marseille, France: European Language Resources Association.

Cohan, A., B. Desmet, A. Yates, L. Soldaini, S. MacAvaney, and N. Goharian. 2018. SMHD: a large-scale resource for exploring online language usage for multiple mental health conditions. In *Proceedings of the 27th International Conference on Computational Linguistics*, pages 1485–1497, Santa Fe, New Mexico, USA: Association for Computational Linguistics.

Conneau, A., K. Khandelwal, N. Goyal, V. Chaudhary, G. Wenzek, F. Guzmán, E. Grave, M. Ott, L. Zettlemoyer, and V. Stoyanov. 2020. Unsupervised cross-lingual representation learning at scale. In *Proceedings of the 58th Annual Meeting of the Association for Computational Linguistics*, pages 8440–8451, Online. Association for Computational Linguistics.

Coppersmith, G., M. Dredze, and C. Harman. 2014. Quantifying mental health signals in Twitter. In *Proceedings of the Workshop on Computational Linguistics and Clinical Psychology: From Linguistic Signal to Clinical Reality*, pages 51–60, Baltimore, Maryland, USA: Association for Computational Linguistics.

Delahunty, F., R. Johansson, and M. Arcan. 2019. Passive diagnosis incorporating the PHQ-4 for depression and anxiety. In *Proceedings of the Fourth Social Media Mining for Health Applications (#SMM4H) Workshop & Shared Task*, pages 40–46, Florence, Italy: Association for Computational Linguistics.

Devlin, J., M.-W. Chang, K. Lee, and K. Toutanova. 2019. BERT: Pre-training of deep bidirectional transformers for language understanding. In *Proceedings of the 2019 Conference of the North American Chapter of the Association for Computational Linguistics: Human Language Technologies, Volume 1 (Long and Short Papers)*, pages 4171–4186, Minneapolis, Minnesota: Association for Computational Linguistics.

Ellendorff, T., S. Foster, and F. Rinaldi. 2016. The PsyMine corpus - a corpus annotated with psychiatric disorders and their etiological factors. In *Proceedings of the Tenth International Conference on Language Resources and Evaluation (LREC'16)*, pages 3723–3729, Portorož, Slovenia: European Language Resources Association (ELRA).

Evans, R., C. Orăsan, and I. Dornescu. 2014. An evaluation of syntactic simplification rules for people with autism. In *Proceedings of the 3rd Workshop on Predicting and Improving Text Readability for Target Reader Populations (PITR)*, pages 131–140, Gothenburg, Sweden: Association for Computational Linguistics.

Fadhil, A., and A. AbuRa'ed. 2019. OlloBot - towards a text-based Arabic health conversational agent: Evaluation and results. In *Proceedings of the International Conference on Recent Advances in Natural Language Processing (RANLP 2019)*, pages 295–303, Varna, Bulgaria: INCOMA Ltd.

Friðriksdóttir, S. R., and A. K. Ingason. 2020. Disambiguating confusion sets as an aid for dyslexic spelling. In *Proceedings of the 1st Workshop on Tools and Resources to Empower People with REAding DIfficulties (READI)*, pages 1–5, Marseille, France: European Language Resources Association.

Gala, N., and J. Ziegler. 2016. Reducing lexical complexity as a tool to increase text accessibility for children with dyslexia. In *Proceedings of the Workshop on Computational Linguistics for Linguistic Complexity (CL4LC)*, pages 59–66, Osaka, Japan: The COLING 2016 Organizing Committee.

Ghosh, S., A. Ekbal, and P. Bhattacharyya. 2020. CEASE, a corpus of emotion annotated suicide notes in English. In *Proceedings of the 12th Language Resources and Evaluation Conference*, pages 1618–1626, Marseille, France: European Language Resources Association.

Gildea, D., M.-Y. Kan, N. Madnani, C. Teichmann, and M. Villalba. 2018. The ACL Anthology: Current state and future directions. In *Proceedings of Workshop for NLP Open Source Software (NLP-OSS)*, pages 23–28, Melbourne, Australia: Association for Computational Linguistics.

Gkotsis, G., A. Oellrich, T. Hubbard, R. Dobson, M. Liakata, S. Velupillai, and R. Dutta. 2016a. The language of mental health problems in social media. In *Proceedings of the Third Workshop on Computational Linguistics and Clinical Psychology*, pages 63–73, San Diego, CA, USA: Association for Computational Linguistics.

Gkotsis, G., S. Velupillai, A. Oellrich, H. Dean, M. Liakata, and R. Dutta. 2016b. Don't let notes be misunderstood: A negation detection method for assessing risk of suicide in mental health records. In *Proceedings of the Third Workshop on Computational Linguistics and Clinical Psychology*, pages 95–105, San Diego, CA, USA: Association for Computational Linguistics.

Gorrell, G., A. Roberts, R. Jackson, and R. Stewart. 2013. Finding negative symptoms of schizophrenia in patient records. In *Proceedings of the Workshop on NLP for Medicine and Biology associated with RANLP 2013*, pages 9–17, Hissar, Bulgaria: INCOMA Ltd.

Graham, S., C. Depp, E. E. Lee, C. Nebeker, X. Tu, H.-C. Kim, and D. V. Jeste. 2019. Artificial intelligence for mental health and mental illnesses: An overview. *Current Psychiatry Reports*, 21, no. 11: 116.

Gratch, J., R. Artstein, G. Lucas, G. Stratou, S. Scherer, A. Nazarian, R. Wood, J. Boberg, D. DeVault, S. Marsella, D. Traum, S. Rizzo, and L.-P. Morency. 2014. The distress analysis interview corpus of human and computer interviews. In *Proceedings of the Ninth International Conference on Language Resources and Evaluation (LREC'14)*, pages 3123–3128, Reykjavik, Iceland: European Language Resources Association (ELRA).

Gutiérrez, E. D., G. Cecchi, C. Corcoran, and P. Corlett. 2017. Using automated metaphor identification to aid in detection and prediction of first-episode schizophrenia. In *Proceedings of the 2017 Conference on Empirical Methods in Natural Language Processing*, pages 2923–2930, Copenhagen, Denmark: Association for Computational Linguistics.

Hauser, M., E. Sariyanidi, B. Tunc, C. Zampella, E. Brodkin, R. Schultz, and J. Parish-Morris. 2019. Using natural conversations to classify autism with limited data: Age matters. In *Proceedings of the Sixth Workshop on Computational Linguistics and Clinical Psychology*, pages 45–54, Minneapolis, Minnesota: Association for Computational Linguistics.

Heeman, P., R. Lunsford, E. Selfridge, L. Black, and J. van Santen. 2010. Autism and interactional aspects of dialogue. In *Proceedings of the SIGDIAL 2010 Conference*, pages 249–252, Tokyo, Japan: Association for Computational Linguistics.

Holmqvist, M. 2013. *Psychological Disorder*, pages 1554–1556. New York: Springer New York.

Howes, C., M. Purver, R. McCabe, P. G. T. Healey, and M. Lavelle. 2012. Predicting adherence to treatment for schizophrenia from dialogue transcripts. In *Proceedings of the 13th Annual Meeting of the Special Interest Group on Discourse and Dialogue*, pages 79–83, Seoul, South Korea: Association for Computational Linguistics.

Husseini Orabi, A., P. Buddhitha, M. Husseini Orabi, and D. Inkpen. 2018. Deep learning for depression detection of Twitter users. In *Proceedings of the Fifth Workshop on Computational Linguistics and Clinical Psychology: From Keyboard to Clinic*, pages 88–97, New Orleans, LA: Association for Computational Linguistics.

Ireland, M., and M. Iserman. 2018. Within and between-person differences in language used across anxiety support and neutral Reddit communities. In *Proceedings of the Fifth Workshop on Computational Linguistics and Clinical Psychology: From Keyboard to Clinic*, pages 182–193, New Orleans, LA: Association for Computational Linguistics.

Jamil, Z., D. Inkpen, P. Buddhitha, and K. White. 2017. Monitoring tweets for depression to detect at-risk users. In *Proceedings of the Fourth Workshop on Computational Linguistics and Clinical Psychology — From Linguistic Signal to Clinical Reality*, pages 32–40, Vancouver, BC: Association for Computational Linguistics.

Ji, Y., H. Hong, R. Arriaga, A. Rozga, G. Abowd, and J. Eisenstein. 2014. Mining themes and interests in the asperger's and autism community. In *Proceedings of the Workshop on Computational Linguistics and Clinical Psychology: From Linguistic Signal to Clinical Reality*, pages 97–106, Baltimore, Maryland, USA: Association for Computational Linguistics.

Jurafsky, D., and C. Manning. 2012. Natural language processing. *Instructor*, 212, no. 998: 3482.

Just, S., E. Haegert, N. Kořánová, A.-L. Bröcker, I. Nenchev, J. Funcke, C. Montag, and M. Stede. 2019. Coherence models in schizophrenia. In *Proceedings of the Sixth Workshop on Computational Linguistics and Clinical Psychology*, pages 126–136, Minneapolis, Minnesota: Association for Computational Linguistics.

Kisely, S. 2020. Unravelling the complexities of inequalities in mental healthcare and outcomes for cultural and linguistic minorities. *BJPsych International*, 17, no. 2: 34–37.

Lala, D., P. Milhorat, K. Inoue, M. Ishida, K. Takanashi, and T. Kawahara. 2017. Attentive listening system with backchanneling, response generation and flexible turn-taking. In *Proceedings of the 18th Annual SIGdial Meeting on Discourse and Dialogue*, pages 127–136, Saarbrücken, Germany: Association for Computational Linguistics.

Lamers, S. M., K. P. Truong, B. Steunenberg, F. de Jong, and G. J. Westerhof. 2014. Applying prosodic speech features in mental health care: An exploratory study in a life- review intervention for depression. In *Proceedings of the Workshop on Computational Linguistics and Clinical Psychology: From Linguistic Signal to Clinical Reality*, pages 61–68, Baltimore, Maryland, USA: Association for Computational Linguistics.

Lee, J., T. Cai, W. Xie, and L. Xing. 2020. A counselling corpus in Cantonese. In *Proceedings of the 1st Joint Workshop on Spoken Language Technologies for Under-resourced languages (SLTU) and Collaboration and Computing for Under-Resourced Languages (CCURL)*, pages 358–361, Marseille, France: European Language Resources association.

Liu, W., and S. Cai. 2015. Translating electronic health record notes from English to Spanish: A preliminary study. In *Proceedings of BioNLP 15*, pages 134–140, Beijing, China: Association for Computational Linguistics.

Loveys, K., P. Crutchley, E. Wyatt, and G. Coppersmith. 2017. Small but mighty: Affective micropatterns for quantifying mental health from social media language. In *Proceedings of the Fourth Workshop on Computational Linguistics and Clinical Psychology — From Linguistic Signal to Clinical Reality*, pages 85–95, Vancouver, BC: Association for Computational Linguistics.

Loveys, K., J. Torrez, A. Fine, G. Moriarty, and G. Coppersmith. 2018. Cross-cultural differences in language markers of depression online. In *Proceedings of the Fifth Workshop on Computational Linguistics and Clinical Psychology: From Keyboard to Clinic*, pages 78–87, New Orleans, LA: Association for Computational Linguistics.

Lynn, V., A. Goodman, K. Niederhoffer, K. Loveys, P. Resnik, and H. A. Schwartz. 2018. CLPsych 2018 shared task: Predicting current and future psychological health from childhood essays. In *Proceedings of the Fifth Workshop on Computational Linguistics and Clinical Psychology: From Keyboard to Clinic*, pages 37–46, New Orleans, LA: Association for Computational Linguistics.

Malin, B. A., K. E. Emam, and C. M. O'Keefe. 2013. Biomedical data privacy: Problems, perspectives, and recent advances. Journal of the American Medical Informatics Association. Volume 20, Issue 1, Pages 2–6.

Matero, M., and H. A. Schwartz. 2020. Autoregressive affective language forecasting: A self-supervised task. In *Proceedings of the 28th International Conference on Computational Linguistics*, pages 2913–2923, Barcelona, Spain (Online): International Committee on Computational Linguistics.

Mayhew, S., C.-T. Tsai, and D. Roth. 2017. Cheap translation for cross-lingual named entity recognition. In *Proceedings of the 2017 Conference on Empirical Methods in Natural Language Processing*, pages 2536–2545, Copenhagen, Denmark: Association for Computational Linguistics.

McDonald, R., S. Petrov, and K. Hall. 2011. Multi-source transfer of delexicalized dependency parsers. In *Proceedings of the 2011 Conference on Empirical Methods in Natural Language Processing*, pages 62–72, Edinburgh, Scotland, UK: Association for Computational Linguistics.

Mianji, F., J. Tomaro, and L. J. Kirmayer. 2020. Linguistic and cultural barriers to access and utilization of mental health care for farsi-speaking newcomers in quebec. *International Journal of Migration, Health and Social Care*, 16, no. 4: 495–510.

Mikal, J., S. Hurst, and M. Conway. 2017. Investigating patient attitudes towards the use of social media data to augment depression diagnosis and treatment: A qualitative study. In *Proceedings of the Fourth Workshop on Computational Linguistics and Clinical Psychology — From Linguistic Signal to Clinical Reality*, pages 41–47, Vancouver, BC: Association for Computational Linguistics.

Milne, D. N., G. Pink, B. Hachey, and R. A. Calvo. 2016. CLPsych 2016 shared task: Triaging content in online peer-support forums. In *Proceedings of the Third Workshop on Computational Linguistics and Clinical Psychology*, pages 118–127, San Diego, CA, USA: Association for Computational Linguistics.

Mitchell, M., K. Hollingshead, and G. Coppersmith. 2015. Quantifying the language of schizophrenia in social media. In *Proceedings of the 2nd Workshop on Computational Linguistics and Clinical Psychology: From Linguistic Signal to Clinical Reality*, pages 11–20, Denver, Colorado. Association for Computational Linguistics.

Mittelstadt, B. 2019. Principles alone cannot guarantee ethical ai. *Nature Machine Intelligence*, 1, no. 11: 501–507.

Mnookin, S. 2016. Out of the shadows: Making mental health a global development priority (English). http://documents.worldbank.org/curated/en/270131468187759113/Out-of-the-shadows-making-mental-health-a-global-development-priority. Accessed February 3, 2021.

Mowery, D. L., A. Park, C. Bryan, and M. Conway. 2016. Towards automatically classifying depressive symptoms from Twitter data for population health. In *Proceedings of the Workshop on Computational Modeling of People's Opinions, Personality, and Emotions in Social Media (PEOPLES)*, pages 182–191, Osaka, Japan: The COLING 2016 Organizing Committee.

Névéol, A., H. Dalianis, S. Velupillai, G. Savova, and P. Zweigenbaum. 2018. Clinical natural language processing in languages other than english: opportunities and challenges. *Journal of Biomedical Semantics*, 9, no. 1: 12.

Parish-Morris, J., E. Sariyanidi, C. Zampella, G. K. Bartley, E. Ferguson, A. A. Pallathra, L. Bateman, S. Plate, M. Cola, J. Pandey, E. S. Brodkin, R. T. Schultz, and B. Tunç. 2018. Oral-motor and lexical diversity during naturalistic conversations in adults with autism spectrum disorder. In *Proceedings of the Fifth Workshop on Computational Linguistics and Clinical Psychology: From Keyboard to Clinic*, pages 147–157, New Orleans, LA: Association for Computational Linguistics.

Park, S., K. Park, J. Ahn, and A. Oh. 2020. Suicidal risk detection for military personnel. In *Proceedings of the 2020 Conference on Empirical Methods in Natural Language Processing (EMNLP)*, pages 2523–2531, Online. Association for Computational Linguistics.

Pfeiffer, J., A. Rücklé, C. Poth, A. Kamath, I. Vulić, S. Ruder, K. Cho, and I. Gurevych. 2020. AdapterHub: A framework for adapting transformers. In *Proceedings of the 2020 Conference on Empirical Methods in Natural Language Processing: System Demonstrations*, pages 46–54, Online. Association for Computational Linguistics.

Phang, J., I. Calixto, P. M. Htut, Y. Pruksachatkun, H. Liu, C. Vania, K. Kann, and S. R. Bowman. 2020. English intermediate task training improves zero-shot cross-lingual transfer too. In *Proceedings of the 1st Conference of the Asia-Pacific Chapter of the Association for Computational Linguistics and the 10th International Joint Conference on Natural Language Processing*, pages 557–575, Suzhou, China: Association for Computational Linguistics.

Pruksachatkun, Y., J. Phang, H. Liu, P. M. Htut, X. Zhang, R. Y. Pang, C. Vania, K. Kann, and S. R. Bowman 2020. Intermediate-task transfer learning with pretrained language models: When and why does it work? In *Proceedings of the 58th Annual Meeting of the Association for Computational Linguistics*, pages 5231–5247, Online. Association for Computational Linguistics.

Quiniou, S., and B. Daille. 2018. Towards a diagnosis of textual difficulties for children with dyslexia. In *Proceedings of the Eleventh International Conference on Language Resources and Evaluation (LREC 2018)*, Miyazaki, Japan: European Language Resources Association (ELRA).

Radford, K., L. Lavrencic, R. Peters, K. Kiely, B. Hachey, S. Nowson, and W. Radford. 2018. Can adult mental health be predicted by childhood future-self narratives? Insights from the CLPsych 2018 shared task. In *Proceedings of the Fifth Workshop on Computational Linguistics and Clinical Psychology: From Keyboard to Clinic*, pages 126–135, New Orleans, LA: Association for Computational Linguistics.

Rauschenberger, M., L. Rello, S. Füchsel, and J. Thomaschewski. 2016. A language resource of German errors written by children with dyslexia. In *Proceedings of the Tenth International Conference on Language Resources and Evaluation (LREC'16)*, pages 83–87, Portorož, Slovenia: European Language Resources Association (ELRA).

Regneri, M., D. King, F. Walji, and O. Palikara. 2020. Images and imagination: Automated analysis of priming effects related to autism spectrum disorder and developmental language disorder. In *Proceedings of the Workshop on Cognitive Modeling and Computational Linguistics*, pages 11–27, Online. Association for Computational Linguistics.

Rello, L., R. Baeza-Yates, and J. Llisterri. 2014a. DysList: An annotated resource of dyslexic errors. In *Proceedings of the Ninth International Conference on Language Resources and Evaluation (LREC'14)*, pages 1289–1296, Reykjavik, Iceland: European Language Resources Association (ELRA).

Rello, L., H. Saggion, and R. Baeza-Yates. 2014b. Keyword highlighting improves comprehension for people with dyslexia. In *Proceedings of the 3rd Workshop on Predicting and Improving Text Readability for Target Reader Populations (PITR)*, pages 30–37, Gothenburg, Sweden: Association for Computational Linguistics.

Rello, L., H. Saggion, R. Baeza-Yates, and E. Graells. 2012. Graphical schemes may improve readability but not understandability for people with dyslexia. In *Proceedings of the First Workshop on Predicting and Improving Text Readability for target reader populations*, pages25–32, Montréal, Canada: Association for Computational Linguistics.

Resnik, P., W. Armstrong, L. Claudino, T. Nguyen, V.-A. Nguyen, and J. Boyd-Graber. 2015. Beyond LDA: Exploring supervised topic modeling for depression-related language in Twitter. In *Proceedings of the 2nd Workshop on Computational Linguistics and Clinical Psychology: From Linguistic Signal to Clinical Reality*, pages 99–107, Denver, Colorado: Association for Computational Linguistics.

Santos, W., A. Funabashi, and I. Paraboni. 2020. Searching Brazilian Twitter for signs of mental health issues. In *Proceedings of the 12th Language Resources and Evaluation Conference*, pages 6111–6117, Marseille, France: European Language Resources Association.

Shing, H.-C., S. Nair, A. Zirikly, M. Friedenberg, Daumé III, H., and P. Resnik. 2018. Expert, crowdsourced, and machine assessment of suicide risk via online postings. In *Proceedings of the Fifth Workshop on Computational Linguistics and Clinical Psychology: From Keyboard to Clinic*, pages 25–36, New Orleans, LA: Association for Computational Linguistics.

Song, H., J. You, J.-W. Chung, and J. C. Park. 2018. Feature attention network: Interpretable depression detection from social media. In *Proceedings of the 32nd Pacific Asia Conference on Language, Information and Computation*, Hong Kong: Association for Computational Linguistics.

Štajner, S., V. Yaneva, R. Mitkov, and S. P. Ponzetto. 2017. Effects of lexical properties on viewing time per word in autistic and neurotypical readers. In *Proceedings of the 12th Workshop on Innovative Use of NLP for Building Educational Applications*, pages 271–281, Copenhagen, Denmark: Association for Computational Linguistics.

Šuster, S., S. Tulkens, and W. Daelemans. 2017. A short review of ethical challenges in clinical natural language processing. In *Proceedings of the First ACL Workshop on Ethics in Natural Language Processing*, pages 80–87, Valencia, Spain: Association for Computational Linguistics.

Tabak, T., and M. Purver. 2020. Temporal mental health dynamics on social media. In *Proceedings of the 1st Workshop on NLP for COVID-19 (Part 2) at EMNLP 2020*, Online. Association for Computational Linguistics.

Tanaka, H., S. Sakti, G. Neubig, T. Toda, and S. Nakamura. 2014. Linguistic and acoustic features for automatic identification of autism spectrum disorders in children's narrative. In *Proceedings of the Workshop on Computational Linguistics and Clinical Psychology: From Linguistic Signal to Clinical Reality*, pages 88–96, Baltimore, Maryland, USA: Association for Computational Linguistics.

Viani, N., L. Yin, J. Kam, A. Alawi, A. Bittar, R. Dutta, R. Patel, R. Stewart, and S. Velupillai. 2018. Time expressions in mental health records for symptom onset extraction. In *Proceedings of the Ninth International Workshop on Health Text Mining and Information Analysis*, pages 183–192, Brussels, Belgium: Association for Computational Linguistics.

Ward, A., and R. Crowley. 2011. Story assembly in a dyslexia fluency tutor. In *Proceedings of the Sixth Workshop on Innovative Use of NLP for Building Educational Applications*, pages 130–135, Portland, Oregon: Association for Computational Linguistics.

Whiteford, H. A., A. J. Ferrari, L. Degenhardt, V. Feigin, and T. Vos. 2015. The global burden of mental, neurological and substance use disorders: An analysis from the global burden of disease study 2010. *PloS one*, 10, no. 2: e0116820–e0116820.

World Health Organization. 2018. Mental health: Strengthening our response. https://www. who.int/news-room/fact-sheets/detail/mental-health-strengthening-our-response. Accessed February 12, 2021.

Yadav, S., J. Chauhan, J. P. Sain, K. Thirunarayan, A. Sheth, and J. Schumm. 2020. Identifying depressive symptoms from tweets: Figurative language enabled multitask learning framework. In *Proceedings of the 28th International Conference on Computational Linguistics*, pages 696–709, Barcelona, Spain (Online): International Committee on Computational Linguistics.

Yaneva, V. 2016. *Assessing text and web accessibility for people with autism spectrum disorder*. PhD thesis, University of Wolverhampton. Wolverhampton, United Kingdom.

Yaneva, V., C. Orăsan, R. Evans, and O. Rohanian. 2017. Combining multiple corpora for readability assessment for people with cognitive disabilities. In *Proceedings of the 12th Workshop on Innovative Use of NLP for Building Educational Applications*, pages 121–132, Copenhagen, Denmark: Association for Computational Linguistics.

Yaneva, V., C. Orasan, L. A. Ha, and N. Ponomareva. 2019. A survey of the perceived text adaptation needs of adults with autism. In *Proceedings of the International Conference on Recent Advances in Natural Language Processing (RANLP 2019)*, pages 1356–1363, Varna, Bulgaria: INCOMA Ltd.

Yaneva, V., I. Temnikova, and R. Mitkov. 2016. A corpus of text data and gaze fixations from autistic and non-autistic adults. In *Proceedings of the Tenth International Conference on Language Resources and Evaluation (LREC'16)*, pages 480–487, Portorož, Slovenia: European Language Resources Association (ELRA).

Yates A., A. Cohan, and N. Goharian. 2017. Depression and self-harm risk assessment in online forums. In *Proceedings of the 2017 Conference on Empirical Methods in Natural Language Processing*, pages 2968–2978, Copenhagen, Denmark: Association for Computational Linguistics.

Yip, J. W. C. 2018. Communicating social support in online self-help groups for anxiety and depression: A qualitative discourse analysis. In *Proceedings of the 32nd Pacific Asia Conference on Language, Information and Computation*, Hong Kong: Association for Computational Linguistics.

Zirikly, A., P. Resnik, Ö. Uzuner, and K. Hollingshead. 2019. CLPsych 2019 shared task: Predicting the degree of suicide risk in Reddit posts. In *Proceedings of the Sixth Workshop on Computational Linguistics and Clinical Psychology*, pages 24–33, Minneapolis, Minnesota: Association for Computational Linguistics.

4 Healthcare NLP Infrastructure for the Greek Language

Aristides Vagelatos
CTI, Computer Technology Institute, Athens, Greece

Elena Mantzari
Neurolingo LP, Athens, Greece

Mavina Pantazara
National and Kapodistrian University of Athens, Athens, Greece

Christos Tsalidis
Neurolingo LP, Athens, Greece

Chryssoula Kalamara
General Hospital of Athens, Athens, Greece

CONTENTS

DOI: 10.1201/9781003138013-4

4.1   INTRODUCTION

The English language is immensely rich in biomedical resources, including an overwhelming amount of medical data, scientific literature, and popular health-related texts on the Web, with standardized field terminologies and ontologies (e.g., LOINC, SNOMED, MED, UMLS) (Dalianis 2018), and advanced domain-dedicated tools, such as biomedical concept extractors and annotators (e.g., cTAKES (Savova et al. 2010), MetaMap (Aronson and Lang 2010), NOBLE Coder (Tseytlin et al. 2016)). Unfortunately, this is not the case for low-resource languages. In a recent review concerning Natural Language Processing (NLP) applications, Névéol et al. (2018) have identified the need for shared tasks and datasets for clinical data processing in languages other than English, which calls for "more publication guidelines that incorporate information about each language other than English and task" to enable the comparison and the collaboration within and across languages.

In this manner, in this chapter, we present a structured description of an NLP resource suite for the Greek language that we have developed within a period of more than ten years and has been deployed in various projects (Vagelatos et al. 2011). The components of the infrastructure are classified into two main categories: (a) those developed de novo to meet the needs of the domain-specific requirements, and (b) existing general language processing tools for Greek that were enhanced or adapted into a robust NLP architecture.

The new assets include: (a) a biomedical corpus of 11.5 million words consisting of a large variety of topics and subtopics, yet restricted to research data rather than clinical ones (Tsalidis et al. 2007), (b) a generic and application-independent medical ontology, based initially on the top-level ontology of Unified Medical Language System (UMLS) (available at https://www.nlm.nih.gov/research/umls/index.html), which mapped into Greek (14,000 concepts) and populated by 16,000 Greek terms drawn upon the MeSH-Hellas (Iatrotek 1997), as well as terms extracted from the corpus.

The main components that were enhanced for the processing of the corpus comprise: (a) the tokenization and sentence splitting tools, (b) the lexicon-based morphosyntactic tagger (Tsalidis et al. 2004), and (c) the multi-word recognition mechanism, a feature-based grammar formalism, which was built on certain morphosyntactic and semantic patterns of Greek multi-word complex biomedical terms.

Envisaged applications of the above NLP suite for the Greek language may include comparative studies in biology, healthcare text analytics, deep learning applications in healthcare data (Yu et al. 2019), interlanguage indexing of medical texts (e.g., Medical Subject Heading – MeSH, https://meshb.nlm.nih.gov/), and text mining tasks, such as the recognition of explicit facts from biomedical literature and document summarization (Miotto et al. 2018). Particularly, during pandemics, embedding biomedical resources for lesser-used languages in NLP applications, such as machine translation, sentiment analysis and surveillance of Internet search terms, may improve multilingual communication and enable better preparation and planning of the healthcare systems for those communities.

The chapter is organized as follows: In section 4.2, we present background information with the focus on the classification of resources and the basic NLP applications that are commonly used in the biomedical domain. In section 4.3, we describe general NLP techniques and tools for the Greek language implemented during research activities and acted as a stepping stone for the development of the healthcare NLP infrastructure. In section 4.4, we present the healthcare NLP infrastructure that we have developed and, to the best of our knowledge, is the only such infrastructure that exists for the Greek language. In section 4.5, we provide an outline for future work that will improve and enhance the existing computational tools and resources. Finally, we conclude with some general observations and suggestions on how to advance healthcare NLP for low-resource and less-widely used languages such as Greek.

4.2 BACKGROUND

The digital content concerning biomedicine and healthcare is exponentially proliferated, while the COVID-19 pandemic has boosted the amount of biomedical data worldwide and in languages other than English, which normally dominates the field. Nowadays, it is estimated that more than 3,000 new research articles are published daily in peer-reviewed journals (Lee et al. 2020). This data can be classified into two main categories (Hersh 2014): (a) *knowledge-based information*, which can be further divided into primary literature, reporting original research and clinical trials in journals, books, reports, etc., and secondary information, which consists of reviews, meta-analyses, also including the popular patient-oriented health information that has recently flooded the Web, and (b) *patient-specific information*, often called clinical data, including the patient's medical record, drug prescriptions, medical referrals, etc.

Whilst the infrastructure for the digitalization, processing, delivery, and accessibility of the knowledge-based information depends largely on the impact of the language in which this information is primarily produced (i.e., English), this is not the case for the patient-specific information, for which the relevant infrastructure rests on the degree of organization and informatization of the national health systems. For example, primary literature and secondary knowledge-based biomedical information written in English have been early enough made available as structured data in bibliographical databases (e.g., MEDLINE via PubMeD, CINAHL, EMBASE). Besides, with the advent of online electronic journals and retro-digitalization of paper-based information, this data has also been accessible via web catalogs and aggregators (e.g., HealthFinder, HON Select, Open Directory, MedlinePlus) or as full-text content. On the other hand, as far as the patient data is concerned, the COVID-19 pandemic showed that some countries, albeit having less-widely used official languages (e.g., Taiwan, South Korea, Uruguay, Vietnam), proved surprisingly efficient in biomedical data collection, sharing and analytics. This was made possible thanks to their robust health infrastructure which was based on innovative digitals tools as well as on the harmonization and structuring of the healthcare data (Horgan et al. 2020).

As we run the "era of data", data can be valuable, as far as it exists for less-used languages, only if it can be used, and the use of data means digitalization, collection, structuring, harmonization, accessibility, and interoperability. NLP tools and techniques can promise multiple benefits to the final users of the above-mentioned data, such as scientists and researchers, clinicians, librarians, database curators, patients/consumers, politicians, and other stakeholders in the healthcare area.

The most important aims of NLP can be summarized in the following (Friedman and Elhadad 2014):

a. NLP advances tasks relating to information extraction (IE), also called *text mining,* through the identification and extraction of similar patterns or structures at the concept level across different texts. For example, IE can enable biosurveillance (i.e., helps understanding the prevalence of a particular epidemic by aggregating conceptually similar information extracted from patient medical records) (Chapman et al. 2004), or support scientific discoveries or comparative research hypothesis by associating pieces of conceptually comparable information among different research papers (Maroto et al. 1997). The most common techniques employed for these tasks may range from less advanced linguistic techniques, such as *named-entity recognition* (NER) (i.e., identification of named entities, such as names of people, places, dates), to more advanced linguistic analysis, such as *term extraction* (i.e., single or multi-word terms that designate the knowledge of a domain and its subdomains), and *factuality detection* in clinical texts, namely the identification of whether a symptom or a diagnosis is valid (e.g., X has a fever), weakened (e.g., X has a possible fever) or negated (e.g., X has no cough) (Dalianis 2018: 69–72 for a concise review).

b. NLP facilitates tasks of accessing information in huge electronic collections of biomedical texts through *information retrieval (IR).* More specifically, IR can benefit from NLP techniques in two ways: (i) it can provide librarians and database curators with tools such as term or topic extractors, thus assisting automatic indexing of articles, (ii) it can improve the recall and accuracy of retrieval and information extraction from texts through the incorporation of ontologies and structured terminologies such as thesauri (i.e., the search for documents including *coronary disease* may return documents that include also the broader term *myocardial ischemia* or, in more specific contexts, the term *coronary aneurysm*) (Diem et al. 2007).

c. NLP can enhance the functionalities of an IR system or the Web by enabling healthcare consumers, professionals, and biomedical researchers to submit queries in natural language, instead of a list of keywords, to obtain information on factual evidence, such as "Does WHO recommend the use of dexamethasone for the treatment of patients with severe and critical COVID-19?" or "What has been used to treat Alzheimer that is not a pharmacological substance?" (Demner-Fushman and Lin 2007; Hristovski et al. 2015).

Above and across all these applications, NLP techniques can ensure high validity, precision, and reliability in the automatically extracted outputs. Probabilistic and statistical approaches such as machine learning (ML) (supervised or unsupervised), even though they can outperform linguistic-based methods of NLP in terms of recall in an enormous volume of data, they can be hardly used for text mining for less-widely used languages such as Greek. This is due to the unavailability of massive textual data, as much for general texts as for specialized domain resources. Whilst hybrid approaches combining NLP techniques with statistical methods seem to be able to bridge the gap between more and less-widely used languages' processing, their implementation is hindered by the unavailability of both NLP integrated systems and massive data for the training of statistical algorithms for those languages (Koutsikakis et al. 2020).

4.3 NLP COMPONENTS FOR THE GREEK LANGUAGE

In this section, at first, we present general facts about the Greek language focusing on the linguistic characteristics that have challenged NLP processing. Then, we discuss general language processing tools, resources, and technologies developed within various projects (Tsalidis et al. 2010) and used as a stepping stone to build upon the healthcare NLP infrastructure described in the next section.

4.3.1 GENERAL FACTS ABOUT THE GREEK LANGUAGE AND NLP SUPPORT

Greek is the official language of Greece, one of the two official languages of Cyprus, as well as one of the 24 official languages of the European Union (EU). Having at least 13 million native language speakers[1] in the 2020 (23rd) edition of Ethnologue, a language reference database published by SIL International (Eberhard et al. 2020), Greek has been ranked at 89th position among the 200 most common languages worldwide. The Greek alphabet consists of 24 letters and two diacritics, one for representing the stress mark (e.g., «ί») and one for the diaeresis, consisting of two dots above a vowel letter and indicating a separate syllable (e.g., «ϊ»). Greek letters are also widely used as international symbols in the terminology of several disciplines, such as mathematics, computer science, chemistry, pharmacology, biology, etc.

Certain linguistic characteristics of the Greek language have challenged computational processing while the development of relevant NLP tools and techniques at certain levels of linguistic representation has proved to be a cumbersome process. Some of them reflected in phonology, morphology, syntax, and vocabulary are related to the "diglossia" between two varieties that coexisted during the 19th and the 20th centuries, namely Katharevousa (the high variety) and Demotic (the low variety), with specific areas of use for each of them, Katharevousa being the language of administration and science. In 1976, Demotic was declared the official language of the Greek state. However, this was not enough to eliminate some of the main difficulties of the Greek language NLP must deal with:

  a. Historical orthography: Certain vowel phonemes are realized with multiple orthographic graphemes: /i/ can be represented by η, ι, υ, ει, οι, or υι; /o/

can be spelled with either o or ω, and /e/ can be spelled with either ε or αι. This has an effect not only on lexemes (e.g., 'λήμμα' (lemma), 'λύμα' (waste), 'λίμα' (nail file) are all pronounced /lima/), but also on inflectional affixes (e.g., 'πολλοί' (many) (masculine plural), 'πολλή' (much) (feminine singular), 'πολύ' (neuter singular) are all pronounced /poli/)). Another common source of spelling errors is the lack of standardization in the transcription of foreign loanwords used in both standard and specialized language (e.g., 'τραίνο/τρένο' (train), 'κορωνοϊός/κορονοϊός' (coronavirus), 'βάκιλλος/βάκιλος' (bacillus)).

b. Morphology and inflection: Greek language is characterized by an extensive morphological structure, especially in the noun and verb inflectional systems (i.e., three genders, two numbers and four cases for nouns and adjectives, and a sheer number of discrete stems and endings for verbs which indicate mood, aspect, voice, tense, and person). For verbs, in particular, there can be different inflectional types bearing the same morphological attributes, albeit used in different registers (e.g., 'εθεωρείτο', 'θεωρούνταν', 'θεωρούτανε' [third person of past imperfective passive voice of the verb 'θεωρώ' (consider)]). Different inflectional paradigms can also indicate differences between everyday life language and terminology (e.g., 'οστό' vs 'οστούν' (bone)).

c. Morphosyntactic ambiguities: These are due to homographs, as in the case of verb inflectional types and the deverbal nouns (e.g., 'λύσεις' [second person perfective aspect of the verb 'λύνω' (solve)] vs 'λύσεις' [plural nominative and accusative case of the noun 'λύση' (solution)]), or in the case of most adverbs (e.g., 'καλά' (well)) vs adjectives in plural ('καλά' (good)).

d. Lexical variations: In specialized language, due to the "diglossia", different word forms can coexist, for example, different lexemes for the same concept in everyday life language and terminology (e.g., 'αφτί/ους' (ear), 'συκώτι/ήπαρ' (liver), 'κοκκινίλα/ερύθημα' (rash)), gender variation (e.g., neuter vs masculine 'νεφρό/νεφρός' (kidney)) or stress variation (e.g., 'εμετός/έμετος' (vomit), 'κοιλιά/κοιλία' (tummy)).

e. Word order and syntax: Greek deploys a subject-verb-object pattern (SVO) in syntactic structures. However, unlike languages like English, the Greek word order is not fixed and the constituents of a clause or of a nominal phrase are characterized by great flexibility. Besides, the subject is often omitted (pro-drop language), a characteristic which is directly associated with the rich inflectional system of Greek (i.e., the verbal forms denote the person and the number of the subject). Although the subject can be easily inferred by the context in human processing and understanding, in NLP applications, however, this can cause major syntactic and semantic ambiguities, as well as parsing problems in regard to coreference and anaphora resolution.

NLP as early as in the late 1980s has been identified as a decisive opportunity for Greek to participate in the European Union linguistic landscape on an equal footing with other more widely used languages. In the light of this policy and by means of

European Union and national funding, multiple tools and resources have been developed for Greek NLP. However, relevant works that survey the area are limited (Gavrilidou et al. 2012; Papantoniou and Tzitzikas 2020). According to the most recent one (Papantoniou and Tzitzikas 2020), there are reported 79 published papers drawn upon both research and grey literature concerning NLP resources and tools for Modern Greek. These papers describe developments for various processing levels, such as phonetics, morphology, syntax, word embeddings, pragmatics, sentiment analysis, and question answering. As it is made obvious by this concise overview, a sheer number of the reported works, albeit innovative in terms of research-based approaches, does not seem to have been incorporated or exploited within a larger application framework. Even more importantly, the resources and tools which are related to health can be counted on the fingers of one hand (Vagelatos et al. 2011; Varlokosta et al. 2016; Zervopoulos et al. 2019).

4.3.2 BASIC NLP COMPONENTS FOR GENERAL LANGUAGE TEXTS

In this section, we provide an overview of the standard methods developed for the Greek language at different levels of text processing: (a) character encoding and tokenization, (b) lemmatization, stemming, spell checking, and morphosyntactic tagging, (c) morphosyntactic disambiguation, syntactical analysis and/or parsing, and d) named entity recognition (NER), term extraction, and sentiment analysis.

4.3.2.1 Preprocessing of Texts

The alphabet and the character set used in Greek texts can create problems in the preprocessing. Additionally, there is no unique alphabet and worst there are many character sets for an alphabet that must be considered. These issues appear mainly when it comes to processing documents deriving from old text collections (e.g., Ancient Greek texts, religious texts, etc.). Mainly, there are two alphabets for the Greek texts, the polytonic and the monotonic. Polytonic alphabet currently represents a small number of contemporary publications and texts, but there exist several texts either from digitization of old books or from religious texts that use a subset of polytonic alphabet. The most common character sets (encodings) that still exist are: (a) Windows 1253, (b) Apple Mac, (c) ELOT 928, and (d) UTF-8. The new context follows the UTF (Unicode) standards (https://www.utf8.com/) and is basically in UTF-8 encoding. There are three additional difficulties to encounter handling Greek characters: (a) the existence of words with mixed Greek and Latin character sets, especially for characters with the same representation (so-called homoglyphs, e.g., A, B, E, H, I, K, M, N, O, P, T, Y, X, o, s, ...), (b) characters from the "Symbol" character set that are used in mathematics, all containing Greek characters, (c) the use of the Latin alphabet instead of the Greek one, an informal writing practice known as "Greeklish writing" frequently used in emails, chats, and blogs, especially by younger people. To deal with these complications, we use custom formalisms that define the locale of each source (collection), as well as algorithms and lexicons. Table 4.1 presents an example code of locale definition for the UTF character set.

TABLE 4.1
Code of locale definition for UTF character set

```
<letter name="Alpha"    utf_code="0x391" attributes="alpha,upper,vowel"/>
<letter name="Beta"     utf_code="0x392" attributes="alpha,upper" />
<letter name="Gamma"    utf_code="0x393" attributes="alpha,upper" />
...
<letter name="alpha"    utf_code="0x3B1"  fcap="Alpha"  acap="Alpha"  attributes="alpha,vowel"/>
<letter name="beta"     utf_code="0x3B2"  fcap="Beta"   acap="Beta"   attributes="alpha"/>
<letter name="gamma"    utf_code="0x3B3"  fcap="Gamma"  acap="Gamma"  attributes="alpha"/>
...
<lclass basic="Alpha"/>    <letters>alphaIAlphaIalphaTIAlphaT    </letters></lclass>
<lclass basic="Beta">      <letters>betaIBeta                 </letters></lclass>
<lclass basic="Gamma"/>    <letters>gammaIGamma               </letters></lclass>
```

4.3.2.2   Segmentation and Tokenization

Text analysis follows a two-phase process. In the first phase, the text is separated into paragraphs, sentences, and words or other tokens, such as the punctuation symbols. In the second phase, the synthesis, the tokens are compiled to syntactic structures that facilitate the recognition of named entities and meanings. Tokenization is usually the first component in every NLP pipeline. The main problem in sentence splitting is the recognition of dot ('.') as a sentence delimiter. Dot is used in many different contexts such as in numbers, technical patterns (e.g., dates), etc. The biggest problem in recognizing dot as a sentence delimiter is the abbreviations, where the dot is also a part of the abbreviated form. There are cases where abbreviations of organizations (e.g., hospitals), products (e.g., drugs), persons (e.g., doctors), etc., consist of two or more words ending in dot and starting with upper case letters. In this case, the recognition of the dot as a separator instead of a part of a word can be early enough cumbersome for the text preprocessing. As far as the Greek language is concerned, there is no secure way to deal with the task of abbreviations recognition. The most effective approach has proved to be the use of techniques that combine dictionaries of abbreviations with heuristic patterns. An additional difficulty in text preprocessing is the recognition of punctuation symbols such as comma (',') which can appear in numerals, but also in exceptional cases as part of a word (i.e., 'ό,τι' (anything)).

4.3.2.3   Word Processing

One important feature in processing textual data, especially from social media platforms and chatbots, is the ability to recognize misspelled words. Spelling errors are divided into two main categories, typographic and orthographic (Hládek et al. 2020). Typographic errors are common in all languages and are usually created by mistake, e.g., wrong letter, missing letter, extra letter, wrong order on a pair of letters. The orthographic errors are usually due to the lack of knowledge of the correct word. The misspelled word sounds or looks very similar to the correct one and the correction process must consider the similarity factor and order of the proposed words accordingly. The ordering of the candidate words is accomplished by a distance function that measures the number of changes between the misspelled and correct words. Candidates with fewer changes must be presented first. "Levenshtein distance" (Levenshtein 1966) is the most well-known distance algorithm used to order a list of words (Wagner, and Fischer 1974). The Levenshtein function measures the number of character changes that need to be applied to a word (e.g., unknown) to convert it to another word (e.g., known from a lexicon). For the Greek language, this method is not satisfactory enough because of the phonetic scheme it uses. Pronunciation is rule-based for Greek words with no exceptions. Although there is a direct mapping from orthography to pronunciation, this mapping is not reflexive, i.e., there are many words that have the same pronunciation. This is because, as mentioned above, equivalent phonemes, albeit with different orthographic spelling, e.g., «ει», «οι», «η», «ι», can have the same pronunciation (/e/). The distance of two words that differ only by these phonemes must not be counted with the numbers of letter changes, i.e., 3 and 2 but better as 1. Thus, a customized version of the Levenshtein function has been adopted, based on phoneme changes to order the words appropriately. Phonemes and their equivalency are defined also in the locale. There are up to five levels of equivalency, with lower level denoting stronger similarity while higher level denoting weaker similarity. In the following code, examples of the 2nd level equivalency are presented:

```
<pclass levels="2">
    <phonemes>
        Eta|Iota|Upsilon|Epsilon,Iota|Epsilon,IotaD|Omikron,Iota|
        Omikron,IotaD|Upsilon,Iota|Upsilon,IotaD|IotaD|UpsilonD
    </phonemes>
</pclass>
<pclass levels="2"><phonemes>Omikron|Omega</phonemes></pclass>
```

Locales can be used with all types of dictionaries since they enable efficient fuzzy matching functionality in dictionary access without the need of extra layer/library/algorithm.

4.3.2.4   Sentence-Based Techniques

Ambiguity is an inherent property of natural languages, as it expresses the level of uncertainty about the meaning of a word, phrase, or sentence, and is present in all phases of content analysis. Word ambiguity can be lexical, syntactic, and semantic. Lexical ambiguity is due to polysemy (i.e., the fact that there is more than one meaning for a word in a natural language). The process to resolve the correct meaning is called "word sense disambiguation" and is presented in advanced NLP pipelines. A subcase is considered the Part of Speech (POS) disambiguation, where the process is referred to as tagging, and the program that accomplishes it is called tagger. Usually, a Greek tagger uses a mixed scheme with three phases (Orphanos and Christodoulakis 1999). In the first phase, the word is looked up in the morphological dictionary and, if it is found and it has a single entry, all morphological attributes are returned. In case no single entry is found, a set of rules work on the context of the word, trying to distinguish the correct one. If a word is unknown, a set of rules examine the suffix and other characteristics of the word, trying to guess its POS. To support this processing, a morphological dictionary has been developed (Tsalidis et al. 2004) with ~100,000 lemmas and 1,200,000 words, containing rich morphological and semantic information for lemmas and inflectional or derivational types. The ambiguity of the words in this dictionary has been categorized as follows: (a) words present in different lemmas ~45,000, (b) words with different POS ~32,000, (c) words that differ in at least one morphological attribute, e.g., number, case, gender ~180,000, (d) words that differ in hyphenation ~650, and (e) words with different morphemic structure ~5,000. POS disambiguation and guessing are carried out with the help of decision trees through examination of the local context, achieving an accuracy of ninety-seven percent (97%) in POS disambiguation and eighty-nine percent (89%) in POS guessing (Orphanos, and Christodoulakis 1999).

4.3.2.5   Semantic Analysis

Named Entity Recognition (NER) as well as term recognition are widely used in many NLP applications, such as classification, semantic indexing and searching, virtual assistants, etc. The task of recognition of the syntactic structure of a phrase or a sentence is complex and nondeterministic. Although for multi-word terms the task is simpler because they follow specific morphosyntactic patterns, nevertheless a grammar formalism is necessary to describe these patterns, as well as a resolution algorithm that will apply the grammar to a text (see also section 4.4.2.5). A context-sensitive grammar formalism has been developed, called Kanon (Vagelatos et al. 2011), along with an efficient resolution algorithm that applies the grammar to the text and recognizes syntactic structures. The resolution algorithm applies a surface parsing technique that permits the recognition of parts of a sentence without requiring the syntax recognition of the whole sentence. The following example is a simplified rule in Kanon for candidate terms:

```
[ARULE="TERMS_1_7", VTEXT="TERM", TTEXT=Concatenate("TTEXT"," ")] =>
\
[LEXY->HasMAttrs ([N])],
[LEXY->HasMAttrs ([N,GEN]), GNC_Agreement(NUMBER,[N,GEN])]{2,5}
/ ;
```

The rule describes a morphosyntactic pattern that appears in multi-word terms. According to this pattern, a term can be a sequence of 2–6 nouns ([N]) where, except the first noun, all others are in genitive case ([GEN]) and must agree on the number (singular, plural). GNC_Agreement(NUMBER,[N,GEN]) expresses the number agreement.

NER and term recognition have been automated with Kanon rules and lexicons expressed in XML, SQL tables, and other formats. Both components have been incorporated in pipelines with higher level rules that recognize events or actions correlating the terms and entities with the event. In biomedical applications, these components have been used for handling the MeSH terms as well as named entities such as persons, organizations, products, dates, etc.

4.3.2.6 Ontologies and Other NLP Apparatuses

Ontology is a term (and a branch also) coming from philosophy and attributed to Aristotle (https://en.wikipedia.org/wiki/Ontology). It expresses the study of common characteristics of entities and concepts like "existence", "being", "becoming" and "reality" (Gruber 1993). In information science, ontology expresses a model that codifies the data and behavior of a system (Gašević et al. 2009). In certain domains, like bioinformatics, ontologies are very important and popular. For the English language, there are standardized field terminologies and ontologies ready to be used in software applications (Rector et al. 2019). For the Greek language, the options are very few and mainly are restricted to printed or electronic documents translated from the English corresponding ones. Such examples are MeSH-Hellas (Iatrotek 1997) and ICD-10 (ICD-10 2010) that are the most known medical resources translated into Greek. To export the contents to a useful electronic format (e.g., XML, JSON), a semiautomatic process was adopted. For our needs, only the data schema, i.e., the closed-world reasoning, defined using the OWL formalism (W3C 2012) and the Protégé tool (Gašević et al. 2009) were implemented, leaving the behavior and open-world reasoning for a future version (see also section 4.4.3).

Modern NLP emerges from the opportunity to better utilize the high volume of open data created on the Web (Eisenstein 2019). Using ML and deep learning technologies, this knowledge is encoded into multidimensional numeric vectors that are used in a plethora of classification applications. For the Greek language, the volume of data that are open and in appropriate format is restricted and cannot lead to efficient models. A combination of rule-based technologies utilizing the existing solid knowledge stored on prefabricated language resources, and ML technologies with auto-adaptation characteristics utilizing the knowledge residing inside texts, has proved to deliver the best results (González-Carvajal, and Garrido-Merchán 2020).

Using the technologies mentioned above, several useful components have been developed which can be used either as stand-alone applications, or in suites that implement complex NLP tasks. Some of these components are presented below:

1. **Stemming & lemmatization:** Stemming is an important function in NLP applications, when the main interest is the meaning/concept and less the nuances of morphology. Stemming is achieved by using a process called "Suffix Stripping", which returns the word stem by removing the end part of the word that bears morphological information (i.e., inflection or derivation). Stemming is not considered an efficient technique to deal with Greek words, since in many cases it produces non-existent words. Therefore, it is used only in cases of unknown words. Lemmatization is a similar function with the exception that the produced word, called lemma, is a correct word. Lemma is usually the headword in electronic or printed dictionaries and is used for citation. Accurate lemmatization for Greek language requires the existence of a morphological lexicon. For the medical sector, two additional problems arise: (a) unknown vocabulary that is not negligible, (b) the mixed morphology (Katharevousa vs Demotic forms) that virtually doubles the size of the words.

2. **Fuzzy Matching:** Spelling checking components for Greek language, English, and French as well as mixed cases. In scientific documents, it is common to have multilingual documents, mainly in English and Greek. It is useful to be able to spell the text using a common speller, especially in environments that do not support language identification and spell switching. Applying multiple locales in lexicons, Levenshtein distance, and phonetic equivalency as mentioned above, all offer acceptable solutions.

3. **Keyword extraction:** This component applies tokenization, lemmatization, and stemming to the words of the documents and the result becomes the input to statistical algorithms, such as TF-IDF (Manning et al. 2008) and BM25 (Robertson et al. 1994). The output is the tagging of the documents in a collection with the most "important" words, i.e., keywords that identify the document in the collection. Keywords are useful in indexing, searching, and classifying documents as presented in the components below.

4. **Related documents:** This component utilizes the "Keyword extraction" and NER components (Nadeau, and Satoshi 2007) to find the attributes

that characterize the document and through them to find related documents in a collection. The component can utilize dictionaries or vocabularies, named entities, terms, and events with a customized weighting scheme to compute the level of relation between documents. The results can be useful for scientific articles on the Web, which together with keywords offer a semantic browsing functionality.

5. **Clustering:** For profiling documents and mining the knowledge therein contained, the clustering component implements known algorithms, such as K-Means (Manning et al. 2008) and Hierarchical (Rokach, and Maimon 2005). These algorithms are unsupervised learning techniques that can be successfully used for clustering collections with medical diagnoses or medicines without needing other resources.

6. **Classification:** Previously tagged datasets can feed the supervised classification algorithms such as Naïve Bayes (Russell, and Norvig 2003) and Binomial and Multinomial Logistic Regression (Agresti 2002). The component can function in two modes: (a) the training mode, which processes the dataset and creates a new model and (b) the classification mode, which is based on the model and classifies unknown documents. Clustering and Classification components are parts of the ML functionality.

7. **Performance acceleration:** In statistical NLP and ML algorithms we normally use numerical representations of words and documents and every function uses a huge number of linear algebra computations. This type of computation is accelerated using specific hardware, such as GPUs and FPGAs. Our environment has been integrated with Tornado (https://e2data. eu/blog/flexible-and-efficient-acceleration-of-natural-language-processing-in-e2data), an experimental and state of the art technology that permits the execution of Java code in GPUs and FPGAs with minimum programming effort achieving big performance gain.

8. **Semantic/ontology searching:** Opinion mining and sentiment analysis are extremely active research areas, due to the increased usage of social media, the production of big data in natural language, and the necessity to utilize this data in order to extract real-time social media users' trends, concerns, or their opinions.

9. **Chatbots:** Since most contemporary applications for a variety of products can be used by smart devices, there is an increased need for friendlier man-machine communication making use of natural language-based interfaces (Agresti 2002). Virtual Assistants are NLP-based components used in a variety of applications. The most known type of Virtual Assistant is the chatbot, which imitates human written or spoken language in many well-described tasks, i.e., information, booking, etc. The basic technology for a chatbot development is the Natural Language Understanding (NLU) module which is similar to a NER component. The module estimates the intent of the user (event or action in NER) and its associated entities.

10. **Word Embeddings:** Word representation with a vector of numbers opens a window to the utilization of modern technologies of machine and deep learning emerging from the application of mathematical sectors,

such as linear algebra, statistics, probability theory, etc. These vectors are created from the processing of large corpora and encoding the semantic characteristics of the words by examining the different context environments they appear. According to Distributional Hypothesis (Harris 1954), two words have similar meanings if they can appear in the same context. The word vectors map the words pointing in a high dimensional space; having these properties means that the Euclidean distance of the points corresponding to two words with similar contexts is small. There are different algorithms having different quality characteristics and usage patterns to produce these vectors, such as Wod2Vec (https://arxiv.org/abs/1301.3781v3), Glove (https://nlp.stanford.edu/pubs/glove.pdf), BERT (https://arxiv.org/abs/1810.04805), etc. Word embeddings are often used as inputs to Neural Nets that train models for classification problems (Mikolov et al. 2013). In our systems, word embeddings are also used inside grammar rules to codify the meaning of a word.

4.4 HEALTHCARE NLP INFRASTRUCTURE

In this section, the Healthcare NLP infrastructure that has been developed for the Greek biomedical language is presented. As it has already been mentioned, it strongly relies on the Greek NLP infrastructure presented in the previous section and, as far as we can tell, it is the only such framework that exists for Greek. Most of its components exist for more than a decade and have been used regularly and mainly for research purposes. Most specifically, in the following, first the construction of a Greek biomedical corpus is described. Then, the NLP tools that were utilized are presented. Finally, the implemented biomedical ontology concludes this section. Other interesting but isolated resources, and an extended NER mechanism developed by other researchers for Greek NLP are also presented.

4.4.1 Greek Biomedical Corpus

For the needs of a national project on medical mining, a Greek biomedical corpus was constructed. Following the basic corpus design methodologies, the core design principles were balance and representativeness. Thus, the necessity to develop a Greek biomedical corpus of written texts, covering a wide variety of subdomains relating to biomedicine, was identified as main requirement.

The obvious next phase was corpus annotation which actually represents the distillation procedure that adds value to the texts. For the annotation process, we utilized several different components based on both existing NLP tools, such as the tokenizer, the sentence splitter, and the morphosyntactic tagger (mentioned in section 4.3.2 above), and those developed de novo, such as a biomedical gazetteer, a multi-word term recognizer, and an ontology-based semantic tagger (more on the new components in the next section).

For the construction of the corpus, digital documents stored on health-related Internet sites were collected. Thus, research and academic institutions' websites were crawled, e.g., Hellenic National Documentation Center, Library of National and Kapodistrian

University of Athens, MedNet Hellas (a local Medical Network), etc. These sites included a pretty exhaustive list of Greek biomedical journals as well as proceedings of national medical conferences. As a result, forty websites were selected with suitable medical documents to be included in the corpus. In total, we finished up with virtually 6,250 documents containing around 11.5 million words (Pantazara et al. 2007).

The collected documents were classified with respect to publication medium (type) and subject (topic). Regarding the type of publication, there were just two options: journal articles (full papers or abstracts) and conference papers. For the topic classification, a suitable scheme was developed by medical experts, based on medical specialties (e.g., radiology, anesthesiology, neurology, pediatrics, etc.). It has to be noted that, although the final corpus consisted of a vast diversity of medical topics, its genre representativeness was still restricted to the research (vs clinical) domain, as it contained mostly scientific articles (70%) as well as conference abstracts (30%).

As described in Tsalidis et al. (2007), the final biomedical corpus was classified regarding the number of documents in each topic/medical specialty, and the results are presented in Table 4.2.

4.4.2 NLP Tools and Components

While designing the corpus, it was made clear that several NLP tools were needed in order (a) to be able to process it, (b) to extend the scope of the entire infrastructure, and (c) to develop the appropriate services for the exploitation of this

TABLE 4.2
Number of documents per medical topic

No. of Documents	Medical Specialty	No. of Documents	Medical Specialty
1,265	Dermatology	78	Neurology
810	Medical Issues (in general)	29	Endocrinology
612	Pathologoanatomy	26	Psychiatry
525	Pneumonology	20	Hematology
454	Cardiology	19	Microbiology
403	Gynaecology – Obstetrics	15	Rheumatology
341	Radiology	14	Social Medicine
324	Paediatrics	14	Nephrology
283	Surgery	12	Anesthesiology
231	Otorhinolaryngology	4	Allergy
163	Urology	4	Genetics
162	Orthopaedics	4	General Medicine
143	Gastroenterology	4	Cytology
137	Ophthalmology	2	Forensic Medicine
104	Neurosurgery		

infrastructure. Fortunately, among the team partners, there was enough experience as well as an already implemented base to build upon (see section 4.3). Thus, a number of tools were created and others were modified and improved in order to establish an infrastructure to support: (a) the conversion and cleansing of the collected documents, (b) the tokenization and sentence splitting, (c) the morpho-syntactic tagging, (d) the identification of medical terms within the biomedical corpus, and (e) the detection of syntactic patterns that could constitute multi-word terms. In this framework, the core mechanism (among the various software parts that deal with the biomedical corpus) is the annotation process. In Figure 4.1, the different tasks of this practice are presented schematically.

4.4.2.1   Document Conversion

The collected medical texts were in two different formats: "pdf" or "html". Thus, a conversion to a common format was required (Common Annotation Structure), in order to be processed uniformly. The common format chosen was "plain text", since on the one hand, this was easy to process and on the other hand, only the textual content was of interest; all the other information like styling and formatting was just noise and had to be cleaned off. For this purpose, two different converters were implemented: an "html-2-txt" one and a "pdf-2-txt" one. These components, although they were based on open software, they demanded adequate modifications so that we could deal with the various Greek character sets and retain as much information as possible regarding the documents' format (e.g., titles, etc.).

4.4.2.2   Tokenization and Sentence Splitting

Following the document conversion, the first step in the text (document) analysis was tokenization, that is, conversion of the input character stream to an output token stream. Tokenization consisted of two different stages. In the first one, the text stream of the input was converted to a token stream, considering white space delimiters as well as certain symbols (comma, period, question mark, etc.). Additionally, during this step, the token orthography was recorded. The term "token orthography" includes both the used character set (e.g., Greek, Latin, or both) as well as the letter case (upper, lower, all caps, first cap). For instance, 'τραύμα' is a Greek-letter-lower-case token while 'Trauma' is an English-letter-first-cap token, and 'S.A.R.S.' is an English-letter-all-cap + middle-dots + ending-dot token, etc. A special peculiarity in Greek biomedical documents proved to be the lack of homogeneity in the input alphabet: several biomedical terms, especially multi-word ones, included both Greek and Latin tokens (e.g., 'σύνδρομο Asperger' (Asperger syndrome), 'βιταμίνη D' (vitamin D), etc.). In the second stage, the token stream was refined passing through a fine-tuning module. During this refinement, certain tokens could further split into two or more tokens depending on their "orthography": e.g., a token ending in a comma, exclamation mark, colon, or semi-colon was split into two tokens, with the punctuation mark being one by itself; a token starting with a quote and ending in a quote was split into three tokens, etc.

Sentence splitting was a second process cycle, during which the output stream from the tokenization was examined so that we could detect tokens that conventionally were used as sentence delimiters, i.e., periods, question marks, exclamation marks, etc.

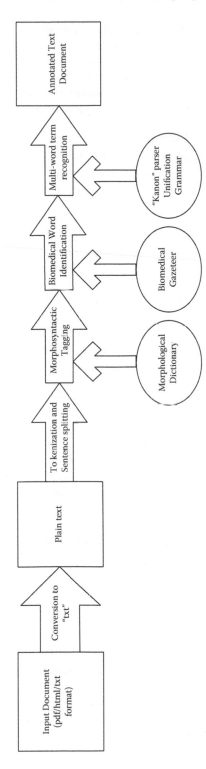

FIGURE 4.1 The tasks of annotation practice: From raw data input to annotated text output.

Particularly tokens ending in a dot were examined thoroughly so that we could resolve whether this dot was (a) part of the token (e.g., in abbreviations) or (b) end of sentence i.e., a sentence delimiter such as full stop. Finally, a very important task was to deal with the abbreviations (with or without dots) which were quite common in medical texts. In order to deal with them, a special dictionary was constructed, containing all the identified abbreviations existing in the biomedical corpus.

4.4.2.3   Morphosyntactic Tagging

A broad-coverage morphological lexicon for the Greek language (more than 100,000 words, almost 1,200,000 word forms) has been developed within previous projects (Tsalidis et al. 2004). This resource has periodically been maintained and has supported NLP applications, especially morphosyntactic tagging. The morphological lexicon is structured in different entries. The basic unit of each entry is lemma, the representative word of all inflectional types of this word. For each lemma, a number of morphosyntactic attributes are codified, such as part of speech, gender, number, and case for nouns, adjectives, articles, pronouns, and voice, tense, mood, number and person for verbs.

Thus, during morphosyntactic tagging, each word was looked-up in the lexicon. There were three possible outcomes: (a) the word was found in one entry of the lexicon, (b) the word was found in two or more morphological entries, and (c) the word was not included in the lexicon. A number of Greek words that appeared in the biomedical corpus could not be found in any morphological entry because most of them were older word forms not used in standard Modern Greek texts (i.e., the word-form 'οστούν' (bone), which occurs in the multi-word term 'ιερό οστούν' (sacrum)).

4.4.2.4   Biomedical Term Identification

After morphosyntactic tagging, the next step was to identify (mark) words that represent biomedical terms. This step is a very important preparatory procedure for the following step (multi-word term recognition). For the recognition of biomedical words, a list (gazetteer) was utilized, which contained about 52,000 biomedical word-forms (virtually 9,000 biomedical terms). This gazetteer was constructed by merging two separate lists: one from the existing morphological lexicon and one that was collected via an automatic or semi-automatic term extraction from the biomedical corpus. As a result, every marked biomedical single-word term candidate could be either a biomedical term in itself or part of a multi-word biomedical term.

4.4.2.5   Multi-word Biomedical Term Recognition

The identification of multi-word biomedical terms proved to be one of the most important and interesting tasks throughout the whole process. In order to be both effective and efficient, the following steps were systematized:

1. First of all, an initial, as well as sufficient body of multi-word terms, was collected. A Greek-English dictionary of biomedical terminology was used as input (Iatrotek 1997).

2. The collected multi-word terms were analyzed and classified according to their structure. Based on this analysis, a number of phrase-structure rules were defined.

3. For the formal expression of these rules, a unification grammar formalism (Shieber 1986) was specified. The following rule is a typical example:

Candidate_Biomedical_Term(Gender1, Number1, Case1) =>
Biomedical_Noun(Gender1, Number1, Case1),
Biomedical_Noun(Gender2, Number2 = singular, Case2 = genitive)

This rule describes the formation of biomedical terms like '*κάταγμα ισχίου*' (hip fractures), '*κακώσεις γονάτου*' (knee injuries), etc.

4. Finally, a parser generator was created. It took as input a unification grammar (written according to the specified formalism) and produced an output that defined the actions to be taken upon rule application. The real parsing was achieved by an execution engine, which analyzed an input document and tried to detect multi-word terms in order to index them following the predefined rules.

4.4.3 GREEK BIOMEDICAL ONTOLOGY CONSTRUCTION

Apart from the corpus and the various tools that were implemented, another major resource that was developed was a Greek biomedical ontology (Vagelatos et al. 2007a). The creation of the ontology was accomplished through two parallel actions: (a) the construction of a "top-level taxonomy" which served as the high-level hierarchy for the knowledge representation of the Greek health care field, and (b) the compilation of the ontology's main structure, that is the choice of the appropriate biomedical terms that instantiate the concepts of the hierarchy nodes.

Regarding the first step, the decision was to adopt an already existing and widely used schema. Thus, after extensive debates the UMLS Semantic Network (SN) (https://semanticnetwork.nlm.nih.gov/) was chosen since mapping the UMLS semantic network into Greek could provide access to the conceptual information for thousands of biomedical concepts included in UMLS. Therefore, the UMLS SN semantic types and relations were translated into Greek.

Having in place the top-level taxonomy, the next step was the enrichment of the taxonomy's nodes. For this task to be fulfilled, two different actions were implemented as described in Figure 4.2:

1. Usage of medical terms that were extracted from several different dictionaries and glossaries (e.g., MeSH-Hellas, terms already existing in the Greek morphological lexicon, etc). These terms were reviewed by experienced linguists as well as medical experts based on both linguistic criteria (i.e., the term was an acceptable Greek word according to the Greek language mechanisms), and domain-specific criteria (i.e., the term designated the appropriate medical concept). This action led to almost 11,000 terms that have been incorporated as instances to the relevant classes of the ontology.

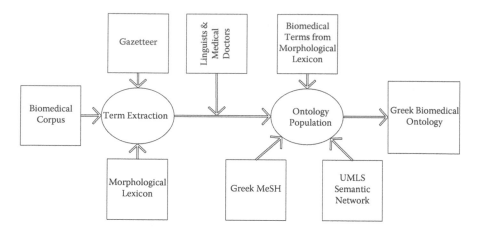

FIGURE 4.2 The ontology construction process.

2. Candidate term extraction from the biomedical corpus (in both automatic and semi-automatic manner). The output was a set of terms that was first assessed by the experts, and then merged into the ontology nodes.

The final version of the produced ontology comprises almost 14,000 concepts, which are represented by 16,500 terms (http://www.iatrolexi.gr:8090/iatrolexi/ontology_browser/index.html). For each entry, the following information was included in the ontology:

- The *concept id*, a unique identifier. This identifier is identical: (a) to the concept id of the UMLS SN for the top-level nodes, (b) to the MeSH-Hellas id for lower concept nodes, while (c) a new and distinct id was assigned to new concepts that were instantiated by Greek terms (acquired from other sources).
- One or more English terms representing that concept.
- One or more corresponding Greek terms representing the same concept.
- A gloss in English.

Two types of relations were encoded for the ontology concepts: the **"is-a" relation** (representing the hierarchical relation, e.g., *DNA packaging* "is-a" type of *Genetic Process*) and the **synonymy relation** (closely related terms that represent the same concept, i.e., the Greek term 'κυτταρική συσσώρευση' is a synonym of the term 'κυτταρική συνάθροιση', both corresponding to the English term 'cell aggregation').

4.4.4 OTHER RELEVANT RESOURCES

Apart from the infrastructure described above, and according to published research as well as grey literature, very few developments have been achieved for the NLP Greek biomedical domain. And most of them are single resources and not integrated environments. Special mention must be made to the following achievements.

4.4.4.1   The Development of ICD-10 for the Greek Language

The adoption of standards and good practices in Greek hospitals, initiated in the first decade of the new millennium, is a reform movement towards the control of medical expenditure mainly in the Hospitals of the Greek NHS (Sarivougioukas and Vagelatos 2020). The new plan included the adoption of the Australian DRG system and the international classification of diseases (ICD-10). In particular, the National School of Public Health translated ICD-10 into Greek (2010) and performed an economic study on the Australian DRGs to localize and adjust their values in the Greek reality. As a result, the Greek translation of ICD-10 is a resource that could be used apart from its main objective for hospital clinical standardization, in other NLP processes as well. On the same line lies the Greek version of the Australian DRGs.

4.4.4.2   Downstream NLP Tasks

On the ML scientific spectrum, a Greek edition of Google's BERT pre-trained language model has been developed (Koutsikakis et al. 2020). According to the authors, the performance evaluation on certain NLP tasks (e.g., text classifications, natural language inference) shows adequate improvement by a margin of 5%–10% over other multilingual Transformer-based models. At the same time, this model has been trained on 29 GB of Greek texts and is hopefully publicly available for research purposes among the interest parties.

Furthermore, a work on NER applied to Greek biomedical texts is described in Tsaknaki and Ioannidou (2017), which although promising, does not seem to have any follow-up regarding its application or integration towards a synthetic intelligent system.

4.5   POSSIBLE APPLICATIONS – FUTURE STEPS

The existence of NLP infrastructure is a prerequisite for the development of adequate computer applications in areas like data/text mining, knowledge-based decision support, automatic translation, terminology management, systems interoperability, integration, etc. A significant body of work that exists today in other languages than Greek reports on experiences with various approaches in crucial areas of research. However, many of these techniques have not yet been tested and evaluated extensively in the biomedical domain, mainly due to the constraints of the medical and biological language (Liu et al. 2011). Wang et al. (2018), in their literature review paper, state the fact that there seems to exist an under-utilization of NLP in clinical records research, which is not easy to understand. The main reason for this fact seems to be that NLP experts have a limited experience in trained data of the clinical domain in comparison to general language data. Additionally, a major drawback lies in the fact that few clinical data sets are available in the public domain.

Nevertheless, in the past few years a breakthrough is going on concerning the various NLP issues and topics addressed by researchers. According to Grouin et al. (2020), 2019 was a very rich period in these scientific areas. It is interesting to stress the fact that, as reported in the abovementioned article, English is by far the first language considered in these studies, with Chinese coming next, French third, and other languages following with just a few references. Again, in this report, the major issue when performing NLP research in the biomedical domain is the access to data. This seems to be the case, considering that the NLP infrastructure is adequate for the languages mentioned above. This fact is also mentioned in Grabar et al. (2019), where it is suggested that:

> Among the languages processed, the main place is occupied by the English language, certainly because the resources (corpora, reference-annotated data, lexica, terminologies, etc.) are mainly available for this language, and also because it is often easier to publish the research work done on the English data.

In this chapter, the main development that has been achieved regarding NLP for the Greek biomedical domain was presented. As it was discussed, Greek is among the so-called less-used languages in the world, regarding first language speakers, 89th in the related table, with almost 13.1 million speakers. On the other hand, the usage statistics of content languages for website, gives us a better position, since Greek is 14th, but with a very small percentage of 0.8%, while English is used by almost 60.5% of the websites (https://w3techs.com/technologies/overview/content_language). These figures support the fact that most of the research on NLP is done on the English language. But it also supports the obvious outcome that languages other than the more-used ones need to develop their own tools in order not to lag far behind the state of the art in these areas.

The infrastructure described here constitutes a considerable innovation for a low-resource language such as Greek. The next steps will be, on the one hand, to improve this infrastructure and, on the other hand, to devise the appropriate applications that can utilize it. Regarding the infrastructure, future work is envisaged on the expansion of term extraction and semantic annotation tools, as well as the enrichment of the linguistic content of the biomedical resources towards bilingualization, which will allow interoperation with NLP biomedical applications of other languages (mainly English), in the perspective of a multilingual biomedical text processing application. Regarding future application development, certain categories can be identified such as: text mining with all its potential applications, data exchange among applications and information integration, comparative studies in biology, indexing of medical texts (e.g., MeSH (Jin et al. 2018)) with special emphasis in interlanguage indexing (e.g., Greek – English), etc. Even without further development, the resources and tools already produced might be useful in various ways. An interesting example is a concordancer (http://www.iatrolexi.gr/iatrolexi/webtools/concordancer/index.html), which started as a support tool for the medical experts during the ontology population (Vagelatos et al. 2007b), and now is freely available for everyone interested in finding usages of terms in context within the collected corpus.

4.6 CONCLUSION

NLP technology is among the technologies that are currently transforming our everyday lives. The way that we interact, work, travel, and shop is not the same anymore. Personal assistants exist that help us in writing better documents, reading a text in a foreign language, and even in communicating with a foreigner that does not speak our language. Of course, there are still weaknesses and pitfalls, for example, in the seamless interaction between a human being and a chatbot, or in our communication with a bot that answers a call to the bank or to our social security organization. Nevertheless, with all these new technological NLP advancements, the future seems brighter and easier even in the most difficult situations. But a question that arises and which seems to be difficult to answer is whether or not, languages that are less used are adequately supported and equally easy to keep the pace that more used languages have already achieved.

In this chapter, we presented the work that has been done towards the support of the Greek language in the biomedical domain regarding NLP technologies and infrastructure. We argued that, although Greek language is near the 90th place in the world's most spoken languages table, an interesting infrastructure has been implemented that one can rely on to develop numerous applications to support the healthcare environment. And it is inevitable that medical experts need these applications to improve their own services.

Nevertheless, more effort needs to be made to improve and expand the tools and resources that have been built till now. Additionally, a better cooperation needs to be established between the different discipline researchers that work in this field: linguists, computer engineers, medical doctors, etc. Moreover, an efficient way should be found for getting together researchers of similar languages that might share acceptable solutions and results. This way less-used languages have a good chance to be part of the revolution that is happening today, in all aspects of NLP, among which the biomedical domain holds a significant place.

Getting involved in biomedical NLP field requires the adaptation to the special type of texts that healthcare deals with. And it is obvious that the number of applications that can be supported is enormous, since healthcare services are vital for every community, ranging from mining medical health data for research purposes to supporting elderly people that receive home healthcare. The only prerequisite is to have the appropriate infrastructure for the language in use.

NOTE

1 Besides, Greek is spoken by around five (5) million people, members of the Greek communities (the Diaspora) who live outside Greece.

REFERENCES

Agresti, A. 2002. *Categorical Data Analysis* (2nd edition). New York: Wiley-Interscience.

Aronson, A., and F.-M. Lang. 2010. An overview of MetaMap: Historical perspective and recent advances. *Journal of the American Medical Informatics Association* 17, no. 3 (May): 229–236. doi:10.1136/jamia.2009.002733

Chapman, W. W., J. N. Dowling, and M. M. Wagner. 2004. Fever detection from free-text clinical records for bio surveillance. *Journal of biomedical informatics* 37, no. 2 (April): 120–127. doi:10.1016/j.jbi.2004.03.002

Dalianis, H. 2018. *Clinical Text Mining: Secondary Use of Electronic Patient Records*. New York: Springer Nature.

Demner-Fushman, D., and J. Lin. 2007. Answering clinical questions with knowledge-based and statistical techniques. *Computational Linguistics* 33, no. 1 (March): 63–103. doi: 10.1162/coli.2007.33.1.63

Diem, L. T. H., J.-P. Chevallet, and D. T. B. Thuy. 2007. Thesaurus-based query and document expansion in conceptual indexing with UMLS: Application in medical information retrieval. In *Proceedings of the IEEE International Conference on Research, Innovation and Vision for the Future, Hanoi, Vietnam, 2007*, 242–246, doi: 10.1109/RIVF.2007.369163

Eberhard, D. M., G. F. Simons, and C. D. Fennig, eds. 2020. *Ethnologue: Languages of the World (23rd edition)*. Dallas, Texas: SIL International. https://www.ethnologue.com/

Eisenstein, J. 2019. *Introduction to Natural Language Processing*. Cambridge, MA: The MIT Press.

Friedman, C., and N. Elhadad. 2014. Natural language processing in health care and biomedicine. In: *Biomedical Informatics: Computer Applications in Health Care and Biomedicine*, ed. E. H. Shortliffe, and J. J. Cimino, 255–284. New York: Springer. doi: 10.1007/978-1-4471-4474-8_8

Gašević, D., D. Djurić, and V. Devedžić. 2009. *Model Driven Engineering and Ontology Development* (2nd edition). New York: Springer.

Gavrilidou, M., M. Koutsombogera, A. Patrikakos, and S. Piperidis. 2012. *The Greek Language in the Digital Age*. New York: Springer. http://www.meta-net.eu/whitepapers/e-book/greek.pdf

González-Carvajal, S., and E. C. Garrido-Merchán. 2020. *Comparing BERT against Traditional Machine Learning Text Classification*. https://arxiv.org/abs/2005.13012

Grabar, N., C. Grouin, and Section Editors for the IMIA Yearbook Section on Natural Language Processing. 2019. A year of papers using biomedical texts: findings from the section on natural language processing of the IMIA yearbook. *Yearbook of Medical Informatics* 28, no. 1: 218–222. doi: 10.1055/s-0039-1677937

Grouin, C., N. Grabar, and Section Editors for the IMIA Yearbook Section on Natural Language Processing. 2020. A year of papers using biomedical texts. *Yearbook of Medical Informatics* 29, no. 1: 221–225. doi: 10.1055/s-0040-1701997

Gruber, T. R. 1993. Toward principles for the design of ontologies used for knowledge sharing. *International Journal of Human-Computer Studies* 43, no. 5-6: 907–928.

Harris, Z. S. 1954. Distributional structure. *Word* 10, no. 2–3: 146–162. doi: 10.1080/0043 7956.1954.11659520

Hersh, W. 2014. Information retrieval and digital libraries. In *Biomedical Informatics: Computer Applications in Health Care and Biomedicine*, ed. E. H. Shortliffe, and J. J. Cimino, 613–642. New York: Springer. doi: 10.1007/978-1-4471-4474-8_21

Hládek, D., J. Staš, and M. Pleva. 2020. Survey of automatic spelling correction. *Electronics* 9, no. 10: 1670. doi: 10.3390/electronics9101670

Horgan, D., J. Hackett, C. B. Westphalen, D. Kalra, E. Richer, M. Romao, A. L. Andreu, J. A. Lal, C. Bernini, B. Tumiene, S. Boccia, and A. Montserrat. 2020. Digitalisation and COVID-19: The perfect storm. *Biomedicine Hub* 5, no. 3 (September–December): 1341–1363. doi: 10.1159/000511232

Hristovski, D., D. Dinevski, A. Kastrin, and T. C. Rindflesch. 2015. Biomedical question answering using semantic relations. *BMC Bioinformatics* 16, article number 6. doi: 10.1186/s12859-014-0365-3

Iatrotek. 1997. *MeSH Hellas. Biomedical Terminology: Greek-English and English-Greek Vocabulary*. Athens: Medical Studies Association.

ICD-10. 2010. International Statistical Classification of Diseases and Related Health Problems. Greek Ministry of Health and Social Solidarity.

Jin, Q., B. Dhingra, W. Cohen, and X. Lu. 2018. Attention MeSH: Simple, effective and interpretable automatic MeSH indexer. In *Proceedings of the 6th BioASQ Workshop A challenge on large-scale biomedical semantic indexing and question answering*,

Brussels, Belgium, November 1st, 2018, 47–56. Brussels: Association for Computational Linguistics. https://www.aclweb.org/anthology/W18-5306

Koutsikakis, J., I. Chalkidis, P. Malakasiotis, and I. Androutsopoulos. 2020. GREEK-BERT: The Greeks visiting Sesame Street. In *Proceedings of the 11th Hellenic Conference on Artificial Intelligence (SETN 2020), Athens, Greece, September 2–4, 2020,* 110–117. New York: Association for Computing Machinery. doi:10.1145/3411408.3411440

Lee, J., W. Yoon, S. Kim, D. Kim, S. Kim, C. H. So, and J. Kang. 2020. BioBERT: A pretrained biomedical language representation model for biomedical text mining. *Bioinformatics* 36, no. 4 (November): 1234–1240. doi:10.1093/bioinformatics/btz682

Levenshtein V. I. 1966. Binary codes capable of correcting deletions, insertions, and reversals. *Soviet Physics Doklady* 36, no. 8 (February): 707–710. Bibcode:1966SPhD... 10..707L.

Liu, K., W. R. Hogan, and R. S. Crowley. 2011. Natural language processing methods and systems for biomedical ontology learning. *Journal of Biomedical Informatics* 44, no. 1 (February): 163–179. doi:10.1016/j.jbi.2010.07.006

Manning, C. D., P. Raghavan, and H. Schutze. 2008. Scoring, term weighting, and the vector space model. In *Introduction to Information Retrieval,* 100–123. Cambridge: Cambridge University Press. doi:10.1017/CBO9780511809071.007

Maroto, M., R. Reshef, A. E. Münsterberg, S. Koester, M. Goulding, and A. B. Lassar. 1997. Ectopic Pax-3 activates MyoD and Myf-5 expression in embryonic mesoderm and neural tissue. *Cell* 89, no. 1 (April): 139–148. doi:10.1016/s0092-8674(00)80190-7

Mikolov, T., I. Sutskever, K. Chen, G. Corrado, and J. Dean. 2013. Distributed representations of words and phrases and their compositionality. https://arxiv.org/abs/1310.4546

Miotto, R., F. Wang, S. Wang, X. Jiang, and J. T. Dudley. 2018. Deep learning for healthcare: Review, opportunities and challenges. *Briefings in Bioinformatics* 19, no. 6 (November): 1236–1246. doi:10.1093/bib/bbx044

Nadeau, D., and S. Satoshi. 2007. A survey of named entity recognition and classification. *Lingvisticae Investigationes* 30: 3–26.

Névéol, A., H. Dalianis, S. Velupillai, G. Savova, and P. Zweigenbaum. 2018. Clinical natural language processing in languages other than English: Opportunities and challenges. *Journal of Biomedical Semantics* 9, article number 12. doi:10.1186/s13326-018-0179-8

Orphanos G., and D. Christodoulakis. 1999. Part-of-speech disambiguation and unknown word guessing with decision trees. In *Proceedings of the 9th EACL Conference, Bergen, Norway, June 8–12, 1999,* 134–141. https://www.aclweb.org/anthology/E99-1018.pdf

Pantazara, M., E. Mantzari, A. Vagelatos, C. Kalamara, and A. Iordanidou. 2007. Development of a Greek biomedical corpus. In *Proceedings of the 11th Panhellenic Conference on Informatics (PCI 2007), Patras, Greece, May 18–20, 2007,* 549–556. Athens, Greece: New Technologies Publications.

Papantoniou, K., and Y. Tzitzikas. 2020. NLP for the Greek language: A brief survey. In *Proceedings of the 11th Hellenic Conference on Artificial Intelligence (SETN 2020), Athens, Greece, September 2–4, 2020,* 101–109. New York: Association for Computing Machinery. doi:10.1145/3411408.3411410

Rector, A., S. Schulz, J. M. Rodrigues, C. G. Chute, and H. Solbrig. 2019. On beyond Gruber: "Ontologies" in today's biomedical information systems and the limits of OWL. *Journal of Biomedical Informatics,* Vol. 100. doi:10.1016/j.yjbinx.2019.100002

Robertson, S. E., S. Walker, S. Jones, M. Hancock-Beaulieu, and M. Gatford. 1994. Okapi at TREC-3. In *Proceedings of the Third Text Retrieval Conference (TREC 1994), Gaithersburg, USA, November 2-4, 1994.* NIST Special Publication 500–225, 109–126. Gaithersburg: National Institute of Standards and Technology (NIST).

Rokach, L., and O. Maimon. 2005. Clustering Methods. In *Data Mining and Knowledge Discovery Handbook*, ed. O. Maimon, and L. Rokach, 321–352. Boston, MA: Springer.

Russell, S., and P. Norvig. 2003 [1995]. *Artificial Intelligence: A Modern Approach* (2nd edition). New Delhi: Prentice-Hall Of India Pvt. Limited.

Sarivougioukas, J., and A. Vagelatos. 2020. Introducing DRGs into Greek national health-care system, in 27 weeks. In *The Importance of Health Informatics in Public Health during a Pandemic*, ed. J. Mantas, A. Hasman, M. S. Househ, P. Gallos, and E. Zoulias, 217–220. Amsterdam, the Netherlands: IOS Press Ebooks. doi:10.3233/SHTI200533

Savova, G., J. J. Masanz, P. V. Ogren, J. Zheng, S. Sohn, K. C. Kipper-Schuler, and C. G. Chute. 2010. Mayo clinical text analysis and knowledge extraction system (cTAKES): Architecture, component evaluation and applications. *Journal of the American Medical Informatics Association* 17, no. 5 (September): 507–513. doi:10.1136/jamia.2009.001560

Shieber, S. M. 1986. *An Introduction to Unification-based Approaches to Grammar*. CSLI lecture notes, no. 4. Stanford: Center for the Study of Language and Information.

Tsaknaki, O., and K. Ioannidou. 2017. A System for named entity recognition and semantic relation extraction: An application to modern Greek biomedical texts. In *Studies in Greek Linguistics 37, Proceedings of the Annual Meeting of the Department of Linguistics, School of Philology, Faculty of Philosophy, Aristotle University of Thessaloniki, May 13–14, 2016* , 713–726. http://ins.web.auth.gr/images/MEG_PLIRI/MEG_37_713_726.pdf

Tsalidis, C., A. Vagelatos and G. Orphanos. 2004. An electronic dictionary as a basis for NLP tools: The Greek case. Paper presented at the *11th Conference on Natural Language Processing (TALN '04), Fez, Morocco, April 19–22, 2004*. https://arxiv.org/abs/cs/0408061

Tsalidis, C., G. Orphanos, E. Mantzari, M. Pantazara, C. Diolis, and A. Vagelatos. 2007. Developing a Greek biomedical corpus towards text mining. In *Proceedings of the 4th Corpus Linguistics Conference, CL2007, Birmingham, UK, July 27–30, 2007*, article number 137. https://www.birmingham.ac.uk/research/activity/corpus/publications/conference-archives/2007-birmingham.aspx

Tsalidis, C., M. Pantazara, P. Minos, and E. Mantzari. 2010. NLP tools for Lexicographic applications in Modern Greek. In *Proceedings of eLexicography in the 21st Century: New Challenges, New Applications (eLex2009), Louvain-la-Neuve, Belgium, October 22–24, 2009*, published in *Cahiers du Cental* 7: 457–462. Louvain: Presses universitaires de Louvain.

Tseytlin E., K. Mitchell, E. Legowski, J. Corrigan, G. Chavan, and R. S. Jacobson. 2016. NOBLE – Flexible concept recognition for large-scale biomedical natural language processing. *BMC Bioinformatics* 17, no 32: 1–15. doi:10.1186/s12859-015-0871-y

Vagelatos, A., E. Mantzari, G. Orphanos, C. Tsalidis, C. Kalamara, and C. Diolis. 2007a. Implementing the NLP Infrastructure for Greek Biomedical Data Mining. Paper presented at the *International Workshop: A Common Natural Language Processing Paradigm for Balkan Languages, RANLP 2007 Conference, Borovets, Bulgaria, September 27–29, 2007*. http://lml.bas.bg/lml/balkan_ws_programme.pdf

Vagelatos, A., E. Mantzari, G. Orphanos, C. Tsalidis, M. Pantazara, C. Kalamara, and C. Diolis. 2007b. Biomedical data mining for the Greek language. In *Proceedings of the 12th World Congress on the Internet in Medicine (MEDNET 2007), Leipzig, Germany, October 7–10, 2007*, published in *Technology and Health-Care* 15, no. 5. Amsterdam: IOS Press.

Vagelatos, A., E. Mantzari, M. Pantazara, C. Tsalidis, and C. Kalamara. 2011. Developing tools and resources for the biomedical domain of the Greek language. *Health Informatics Journal* 17, no. 2 (June): 127–139. doi:10.1177/1460458211405007

Varlokosta, S., S. Stamouli, A. Karasimos, G. Markopoulos, M. Kakavoulia, M. Nerantzini, A. Pantoula, V. Fyndanis, A. Economou, and A. Protopapas. 2016. A Greek corpus of aphasic discourse: Collection, transcription, and annotation specifications. In *Proceedings of LREC 2016 Workshop Resources and Processing of Linguistic and Extra-Linguistic Data from People with Various Forms of Cognitive/Psychiatric Impairments (RaPID-2016), Portorož, Slovenia, May 23–28, 2016*, 14–21. https:// aphasia.talkbank.org/publications/2016/Varlokosta16.pdf

Wagner, R. A., and M. J. Fischer. 1974. The string-to-string correction problem. *Journal of the ACM* 21, no. 1: 168–173. doi:10.1145/321796.321811

Wang, Y., L. Wang, M. Rastegar-Mojarad, S. Moon, F. Shen, N. Afzal, S. Liu, Y. Zeng, S. Mehrabi, S. Sohn, and H. Liu. 2018. Clinical information extraction applications: A literature review. *Journal of Biomedical Informatics* 77 (January): 34–49. doi:10.1016/ j.jbi.2017.11.011

W3C (World Wide Web Consortium) 2012. *OWL 2 Web Ontology Language Document Overview* (2nd edition). https://www.w3.org/TR/2012/REC-owl2-profiles-20121211/

Yu, Y., M. Li, L. Liu, Y. Li, and J. Wang. 2019. Clinical big data and deep learning: Applications, challenges, and future outlooks. *Big Data Mining and Analytics* 2, no. 4 (December): 288–305. doi:10.26599/BDMA.2019.9020007

Zervopoulos, A. D., E. Geramanis, A. Toulakis, A. Papamichail, D. Triantafylloy, T. Tasoulas, and K. Kermanidis. 2019. Language processing for predicting suicidal tendencies: A case study in Greek poetry. In *Proceedings of the 15th IFIP International Conference on Artificial Intelligence Applications and Innovations (AIAI), Hersonissos, Crete, Greece, May 24–26, 2019*, 173–183. New York: Springer.

5 Formalizing the Recognition of Medical Domain Multiword Units

Kristina Kocijan

Department of Information and Communication Sciences,
Faculty of Humanities and Social Sciences, University of
Zagreb, Zagreb, Croatia

Krešimir Šojat

Department of Linguistics, Faculty of Humanities and Social
Sciences, University of Zagreb

CONTENTS

DOI: 10.1201/9781003138013-5

5.1 INTRODUCTION

Multiword units (MWUs) have been the focus of research of many authors since even before the Natural Language Processing (NLP) era, which has only helped to spread interest in MWUs in multiple dimensions and directions. This is also true among computational linguists and other scholars of language whose work has been dedicated to the Croatian language in both its standard and spoken varieties (Frančić and Menac-Mihalić 2013). MWUs in Croatian have been observed from a variety of perspectives, including translation and language comparison (Vidović Bolt 2018; Parizoska 2018), language teaching (Jelčić 2014; Barčot and Milčić 2019), conceptual organization in e-lexicography (Filipović Petrović and Parizoska 2019a; 2019b), domain-specific MWUs – e.g., *marketing* (Rajh 2015), *police* (Radek and Pešut 2016), and *marine* (Borucinsky and Kegalj 2017). More recent corpus-based research has been conducted on Riznica, the Croatian language corpus available via the Institute of Croatian Language and Linguistics (Kovačević and Ramadanović 2013), and on the hrWaC corpus, available through SketchEngine (Rajh 2015; Filipović Petrović and Parizoska 2019b), but in the literature we find some additional corpora built for specific research questions (Agić and Bekavac 2013; Kocijan and Librenjak 2016).

Computational identification of different types of MWU in the Croatian language has been attempted by several reported projects (Tadić and Šojat 2003; Ljubešić et al. 2014; Almić and Šnajder 2014; Kocijan and Librenjak 2016; Šojat et al 2016; Kocijan and Librenjak 2018). However, to our knowledge, this study is the first such attempt on a Croatian medical domain corpus.

NLP is not new to the medical domain, and its importance has been pointed out in papers by both medical and NLP researchers ever since the 1990s (Haug et al. 1990; Friedman and Hripcsak 1998; Meystre and Haug 2006; Doan et al. 2010; Friedman et al. 2013; El-Rab et al. 2017). A short historic overview of the field, including its beginnings in the 1960s and the increase in publications in the domain of medical NLP, was provided by Friedman et al. (2013). The term medical language processing (MLP) is used (Friedman et al. 1998; Barrows et al. 2000) to indicate the ever-stronger connection between NLP and systems developed specifically to detect and annotate information found in medical records so that they can be fed into different applications utilized for decision support, modification of guidelines, process management, therapy and symptom comparison, and other purposes.

After having dealt with single words – namely, domain-specific nouns and adjectives (cf. Kocijan et al. 2020; 2021) – extracted from the Croatian medical corpus, which contains pharmaceutical instructions for medicaments, we are extending our research to multiword units. More precisely, this chapter will tackle the problem of recognizing multiword units (MWUs) in medical domain texts written in the Croatian language. Our focus in this chapter is on the automatic recognition of complex MWUs using low-resource settings. The units we are dealing with here are complex, in that they consist of two or more noun or prepositional phrases, such as 'symptomatic treatment of patients' [*simptomatsko*$^{(A)}$ *liječenje*$^{(N\ NOM)}$ *bolesnika*$^{(N\ GEN)}$] *or* 'herbal anti-asthmatic syrup' [*biljni*$^{(A)}$ *sirup*$^{(N\ NOM)}$ *protiv*$^{(PREP)}$ *kašlja*$^{(N\ GEN)}$, as well as more complex ones, such as 'continuous evaluation of the risk-benefit balance of the drug' [*kontinuirano*[(A)] *praćenje*[(NOUN)] *omjera*[(N GEN)] *koristi*[(N GEN)]*i*[(C)] *rizika*[(N GEN)] *lijeka*[(N GEN)]].

Furthermore, the detection of MWU *boundaries* is important for different text-mining tasks, such as information extraction (Zhu et al. 2013), terminology extraction (Tadić and Šojat 2003; Bekavac and Tadić 2009), or named entities recognition (Agić and Bekavac 2013). If correctly marked, they can contribute significantly to applications used for medical research purposes, decision-making in medical institutions (Friedman et al 2013; Spasić et al. 2014), any type of information-retrieval task, biomedical text mining (including term extraction), or machine translation (Moreno Sandoval et al. 2018). For example, if we are looking for documents on 'adjustment of a drug dose' (*prilagodba doze lijeka*), it is important to differentiate between the text talking about '*potentially needed* adjustment of a drug dose' ('*potencijalno potrebna prilagodba doze lijeka*') and '*non-recommended* adjustment of a drug dose' ('*ne preporuča se prilagodba doze lijeka*').

The research questions we explore here are:

a. Can different linguistic patterns be used to automatically find the boundaries of multiword units in the medical domain?
b. What types of errors can decrease the precision of such an algorithm?
c. Is the length of an MWU project-dependent?

In the following pages of this chapter, we will describe a computer algorithm designed in the NooJ NLP environment, used also in various other NLP tasks (Silberztein 2016), specifically with the purpose of detecting and annotating three different categories of multiword units in texts characteristic of the medical domain. A short overview of previous work in this area will be followed by sections on Croatian language features, corpus description, dictionary data, and algorithm description. A section on the results of the experiment, error analysis and some future objectives will follow.

5.2 RELATED THEORETICAL OVERVIEW

An overview of rule-based approaches to different levels of analysis of medical-related texts, ranging from simple regular expressions to commercial healthcare-domain-oriented tools like ClearForest, LEXIMER, and AeroText, among others, is given in Spasić et al. (2014). A more up-to-date list of biomedical informatics tools is available on the ORBIT website, where 95 tools have been reported so far for different domains, including *relationship recognition, bioinformatics, classification, cluster analysis, text mining, information extraction, and NER*, among others. These tools vary across different medical sub-domains, like RADA, which is used in radiology reports (Johnson et al. 1997), or a specially designed application (using GATE as its base toolkit) in psychology, where a rule-based approach was added to the ML approach in order to improve the program's performance, specially designed to detect negative symptoms of schizophrenia (Gorrell et al. 2013).

A comparison of GDP (Glaucoma Dedicated Parser) and MedLEE (Medical Language Extraction and Encoding) for IE (age, gender, intraocular pressure, cup:disk ratio, visual fields, retinal vascular status, and glaucoma diagnosis) in ophtalmologists' coded notes on glaucoma patients was reported by Barrows et al. (2000). Although MedLEE was designed to process regular grammatically well-formed sentences, it performed better precision-wise in this comparison. A rule-based approach was also used for building the GGPONC corpus of German medical text in oncology, to be more precise in the detection of TNM expressions used to classify malignant tumors (Borchert et al. 2020).

An overview of the rule-based approach used specifically for the extraction of named entities in clinical text (MedLEE, in use since 1994; CLARIT, 1996; SymTex, 1997; MetaMap, 2001; two methods developed in 2007 by Chhieng's team and Levin's team to detect drug names in clinical records and anesthesia records, respectively; MedEx, for medication-related information extraction, 2010) can be found in (Doan and Xu 2010), where they also report on improving the results of their Support Vector Machine when features generated from a rule-based system are included (P: 0.932; R: 0.871; f: 0.900).

MedEx uses context-free grammars to detect pre-defined semantic patterns and scored the second-best results at the 2009 i2b2 NLP challenge (f-measure of 0.821 at the system level and 0.810 at the patient level) (Doan et al. 2010). Similar improvements in two biomedical concept normalization systems – MetaMap and Peregrine – were reported by Kang et al. (2013) when they added a rule-based NLP module to them. The f-measures for MetaMap and Peregrine respectively increased from 0.610 to 0.733 and from 0.639 to 0.780 for boundary matching, and respectively from 0.551 to 0.855 and from 0.569 to 0.736 for concept identifier matching. A Java version of MetaMap (MMTx) was used to build an automated problem list solver in the area of cardiovascular medicine and surgery (Meystre and Haug 2006). The authors reported a recall of 0.896 on their custom-made corpus, which was higher than the results obtained on the original data that MetaMap was trained on.

Mondal et al. (2017) present a rule-based model for extracting eight types of relations in medical text: Drug-SymptomDisease (*Warfarin ... clots ... strokes or heart-attacks*); Anatomy-SymptomDisease (*painful inflammation ... big toe and foot*); Disease-Symptom (*Anal fissures ... pain and bleeding with bowel movements*); Drug-Drug (*oral lipid-lowering medicine (trade name Zocor) ...*); Symptom-Symptom (*rhythmic tightening* in *labor of the upper uterine musculature ...*); Disease-Disease (*Kaposi's sarcoma* is a form of *skin cancer ...*); Miscellaneous Medical Term – SymptomDisease (*... toxic products ... life-threatening symptoms*); Anatomy-Anatomy (*... head* or *body*). The performance of this model was tested against the feature-oriented model, and it scored better in recall (in each separate category and overall), but the f-measure favored the second approach due to lower precision. The NooJ linguistic environment has a very successful application in the medical domain for a variety of tasks, such as the analysis of narrative psychological content (Ehmann and Garami 2010) or the detection of medical named entities in Arabic texts (Boujelben et al. 2011) and in Italian medical texts (di Buono et al. 2015), among others.

The wide usage of rule-based approaches in the medical domain can be explained by the terminology-heavy text type that a rule-based approach handles better (Doan and Xu 2010; Koleck et al 2019; Pilan et al. 2020) but also by the explanatory nature of such rules that enables practitioners to understand what is it that they do (Silberztein 2016). This is contrary to what is usually referred to as the "black box" characteristic of the machine-learning approach. A more recent paper (Wang et al. 2018) gives a review of 263 studies on rule-based vs machine learning methods used within the clinical information extraction (published from 2009 to 2016) and reports detecting 65% rule-based systems among them.

Imaichi et al. (2013) suggest that different approaches to information extraction (ad-hoc rule-based learning vs. machine learning) also depend on the specific task. Their comparison of these two methods applied to Japanese texts shows that the rule-based approach performs better in de-identification jobs (namely, in the detection of age, time, hospital, and gender) while machine learning results in better recall and overall scores for detecting complaints and diagnosis (family member, negation, suspicion, recognition, improvement). Still, numerous examples show that hybrid solutions (i.e., the integration of different methods in a synergistic manner) result in the best performance (Doan and Xu, 2010; Friedman et al. 2013; Kang et al. 2013; Moreno Sandoval et al. 2018).

5.3 LANGUAGE FEATURES

Our main objective is to recognize complex MWUs (i.e., those consisting of two or more noun and/or prepositional phrases), to manually analyze our results and determine how they can be used in further phases of the project. Before we present the results, we shall provide a brief theoretical overview of complex MWUs serving as candidates for medical terms in Croatian. Each noun phrase (NP) in a complex unit contains a noun as the obligatory head and one or more optional adjectives functioning as its modifiers or attributes – see, e.g., (5.1).

$$< tvrda^{(A)} \, kapsula_{NP}^{(N \, NOM)} > - <\text{hard}^{(A)} \text{ capsule}^{(N)}> \qquad (5.1)$$

5.3.1 NOUN PHRASES

As in other Slavic languages, Croatian morphology is very rich in terms of both inflection and word formation. Inflection in Croatian is almost exclusively based on suffixation (Silić and Pranjković 2005). Generally, inflectional morphemes are bound morphemes expressing several grammatical categories; for example, gender, number, and case for nouns and adjectives; tense, person, and number for verbs (Marković 2012). In terms of grammatical gender, Croatian nouns can be masculine, feminine, or neuter. There are two numbers (singular and plural, in rare cases, singulare or plurale tantum) and seven cases. Within noun phrases consisting of nouns and adjectives or other determiners, there is always an agreement in gender, number, and case. NP heads (nouns) determine the morphological form of adjectives and other co-occurring elements such as pronouns, etc. For example, in the MWU *bubrežna*[A] *funkcija*[N NOM] – *kidney function*, the grammatical categories of the adjective *bubrežna* are determined by the categories of the noun *funkcija*. In this example, the categories of nominative, feminine and singular are cumulatively expressed by the suffix – *a* in both members of the NP. Semantically, adjectives in such noun phrases are premodifiers used to specify the meaning of co-occurring nouns.

However, noun phrases can be followed by other noun phrases or prepositional phrases. In complex MWUs, consisting of two or more NPs, noun phrases are used as postmodifiers of preceding NPs. Such post-modifications are in oblique cases, mostly genitive – e.g., (5.2) – or instrumental – e.g., (5.3).

$$simptomatsko^A \ liječenje^{(N \ NOM)} \ bolesnika^{(N \ GEN)}$$

– symptomatic treatment of patients (5.2)

$$trovanje^{(N \ NOM)} \ gljivama^{(N \ INS)} - \text{mushroom poisoning} \qquad (5.3)$$

In some cases, medical terms in Croatian can consist of two noun phrases, both in nominative cases – e.g., (5.4).

$$mala^{(A)} \ žlijezda^{(N \ NOM)} \ slinovnica^{(N \ NOM)} - \text{minor salivary gland} \qquad (5.4)$$

5.3.2 PREPOSITIONAL PHRASES

Noun phrases can be followed and postmodified by prepositional phrases (PPs), as well. Prepositions as heads determine or 'govern' the case of nouns in PPs. The grammatical cases of nouns in prepositional phrases in Croatian are genitive, dative, accusative, locative, and instrumental. Nouns in prepositional phrases can also be optionally pre-modified by one or more adjectives – e.g., (5.5).

$$< tvrda^{(A)} \ kapsula_{NP}^{(N \ NOM)} > \ <s^{(PREP)} \ prilagođenim^{(A)} \ oslobađanjem_{PP}^{(N \ INS)} >$$

–hard capsule with modified release (5.5)

Semantically, prepositional phrases serve as attributes or objects of preceding NPs – e.g., (5.6).

$$< biljni^{(A)} \ sirup^{(N \ NOM)} > \ <protiv^{(PREP)} \ kašlja^{(N \ GEN)} >$$

–herbal antiasthmatic syrup (literally: <herbal syrup$_{NP}$ >

<against cough$_{PP}$>) (5.6)

5.4 CORPUS

5.4.1 RELATED CORPORA

Health care is abundant in free-form medical texts, which are also almost impossible to obtain, even for research purposes. This can, quite understandably, be attributed to confidentiality reasons regardless of the de-identification or anonymization solutions. Although we find descriptions in the literature of patient-related corpora such as ophthalmology visit notes from the NYP clinical data repository (Barrows et al. 2000); digestive surgery patient records from a Geneva hospital (Alphonse and Bouillon 2004); 160 clinical documents in the area of cardiovascular medicine (Meystre and Haug 2006); and 21 000 chest x-ray reports that were added to public datasets from the same domain (Lamare et al. 2020), this obstacle seems to be common among researchers regardless of the language, (cf. for Norwegian – Pilan et al. 2020; for German – Borchert et al. 2020), including English (cf. Friedman et al. 2013).

At the same time, there are plenty of corpora compiled from scientific papers or abstracts. As an illustration, we can mention the *AZDC* from 2009, with 2 784 sentences from Medline abstracts (Kang et al. 2013); the *Harrison* sub-collection of full texts in Spanish; abstracts from five medical journals in Japanese; and full texts from health-related journals in Arabic of MultiMedica corpus (Moreno Sandoval et al. 2018). However, while different platforms for hosting papers on relevant research dominate in texts available in English, like PubMed (Spasić et al. 2014) or specialized health-related social networks like DailyStrength (Leaman et al. 2010), the text available in Croatian is still sparse and so far does not provide enough relevant data for this type of research.

It was for these reasons that we have opted for another source that is rich in medicine-related data but, at the same time, does not have to meet the confidentiality criteria. Such are the drug leaflets with *information on medicaments* authorized for usage in Croatia. The collected corpus thus consists of descriptions of a product (including its composition, usage, side-effects, etc.) prepared in the pdf file format and written in Croatian.

5.4.2 PREPARING THE CROATIAN MEDICAL CORPUS

The corpus compilation routine in NooJ accepts pdf files and converts them to simple text format or **.not* files, characteristic of its working environment. However, during the process, the text inside the table format is converted to a regular text that produces some noise in the semantic analysis. This was noticed during the data analysis of the first results, and such occurrences were manually corrected – i.e., each cell data was followed by a new line feed to separate them. The same procedure was executed for the members of ordered and unordered lists.

Another obstacle to clean data was *stamps* that were added to each page of the document in order to mark their authenticity. Each stamp consists of a word – followed by a date and additional string. The conversion tool in NooJ falsely recognized them as part of the text, inserting in some cases the stamp content in the

TABLE 5.1

Overview of the dataset

Sub-section	No. of documents	No. of tokens
A	10	44 784
H	11	50 253
M	24	174 154
S	17	60 854
V	11	47 526
Total	**73**	**377 571**

middle of a sentence. Nevertheless, we did not make any changes to these insertions, which resulted in some MWUs not being recognized in full or at all.

During the conversion from pdf, there was also a negative transfer of certain letters (mainly letters with diacritics, such as Ž, č, and đ), which were replaced with a standard *find-replace* procedure. Some minor spelling errors were observed, but they were not treated in any way, and no other changes were performed on the corpus.

For this specific research, a sub-corpus with 377 571 tokens was selected from a larger corpus. Table 5.1 provides an overview of the number of documents (73 in total) and the number of tokens in each sub-section (sub-sections: A, H, M, S, V) used in this research. The sub-section names correspond to the first letter of the drug's ATC code (Anatomical Therapeutic Chemical Classification System) to allow for possible future multilingual comparisons.

5.5   DATA FROM THE DICTIONARY

Thanks to the previous work on this corpus (Kocijan et al. 2020; 2021), the texts have already been linguistically annotated. More precisely, all the nouns in the corpus have part-of-speech, type, case, gender, and number tags that were extracted from the NooJ dictionary. The same is true of the adjectives, which are tagged for degree of comparison (*neutral, comparative, superlative*) as well.

Additionally, both nouns and adjectives are enriched with semantic tags <+MED> if associated with the medical domain as shown in (Kocijan et al. 2020; 2021), respectively. The nouns are also semantically categorized in the dictionary mostly with only one medical sub-domain (anatomy, disease, medicine, chemistry, etc.). However, the sub-domain tags were not used in this phase of the research and will not be further discussed at this time.

In the subsequent phase, we dealt with terms consisting of nouns and one or more adjectives serving as attributes. We were able to annotate domain-specific attributes by using already marked nouns from the previous step. We also recognized noun phrases (NPs), which belong to the medical domain either due to the meaning of an adjective – e.g., (5.7) – and/or of a noun – e.g., (5.8).

$$bubrežna^{(A)} funkcija^{(N\ NOM)} - \textbf{kidney} \text{ function} \qquad (5.7)$$

$$akutna^{(A)} \textbf{bol}^{(N\ NOM)} - \text{acute } \textbf{pain} \qquad (5.8)$$

In such cases where the meaning of neither noun nor adjective is specific to the medical domain, the whole MWU is marked and stored as a medical term directly in the dictionary – e.g., (5.9).[1]

$$vodene^{(A)} kozice^{(N\ NOM)} - \text{chickenpox (literally: water}^{(ATTR)} \text{ goat}^{(N\ NOM\_PL\_DIM)})$$
$$(5.9)$$

During the linguistic analysis, simple pattern matching rules in NooJ have been utilized to disambiguate prepositions, nouns, and adjectives, in order to lower the noise among the annotations. Due to the inner and outer ambiguity, mostly due to homography, Croatian words may have a number of different annotations that puts them in several POS categories simultaneously (e.g., *tih* can be both the adjective 'silent' and the pronoun 'those'; *oko* can be the noun 'eye' and the preposition 'around'; *K* can be the preposition 'towards' and the abbreviation for vitamin K, or the chemical element Potassium). This pre-processing phase has helped raise the precision of all three algorithms that we will explain in more detail in the following section.

5.6 ALGORITHM

The importance of detailed and reliable data in the health domain (Chute et al. 1998; Wilcox and Hripcsak 2003) is probably one of the greatest across the disciplines. It is thus essential to learn about data from both the informational and the linguistic perspectives to be efficient and effective. In this era of easily available unstructured data, seen as a *chaos-data* from the perspective of a relational database, NLP tools have proved to be a real ally in bringing order to that chaos (Wilcox and Hripcsak 1998).

With that thought in mind, we have designed an algorithm that can help find candidate terms for semi-automatic terminology dictionaries and detect patterns in naming schemas used for different diseases. This will further allow us not only to recognize the boundaries of multiword units, but also to make relations between an organ and a disease that can additionally be used for automatic mapping of unstructured text to structured databases.

5.6.1 Length of an MWU

There is no consensus on the right length of an MWU; it certainly depends not only on the domain of the text but also on the final purpose of its detection. For example, named entity recognition (NER) wants to find the boundaries for a single drug, disease, or symptom. It is similar in the domain of terminological variation detection (Kang et al. 2013), the aim of which is to recognize and connect different ways of referring to a single term. It would thus be better to keep the boundaries on the smallest MWU – e.g., (5.10).

$oralna^{(A)} primjena^{(N\ NOM)} ibuprofena^{(N\ GEN)}$ –oral administration of ibuprofen

(5.10)

However, more complex tasks, such as information extraction, might give better results if they can detect the longest MWU – e.g., (5.11).

$opsežna^{(A)} toksikološka^{(A)} ispitivanja^{(N\ NOM)} oralne^{(A)} primjene^{(N\ GEN)} ibuprofena^{(N\ GEN)}$

– extensive toxicological studies of oral administration of ibuprofen   (5.11)

The length of complex terms in some medical terminological databases used in various medical language projects, such as SYNTEX (Alphonse and Bouillon 2004), has been reported to be usually between two and five. Furthermore, the same authors reveal that in most cases doctors consider complex terms constructed of three or more words as bad. They provide an example term, *adénocarcinome occlusive de colon transverse,* which doctors propose should be split into two terms, *adénocarcinome occlusive* and *colon transverse,* but also a counterexample, *ablation de lame,* that they suggested should be extended to *ablation de lame ondulée* (Alphonse and Bouillon 2004). Similarly, Moreno Sandoval et al. (2018) give examples of acceptable and not acceptable candidates for multiword terms. They propose that the ATR (Automatic Term Recognizer) 'should detect as candidate terms *infarto* (infarct or heart attack), *infarto de miocardio* (myocardial infarct) and *infarto agudo de miocardio* (acute myocardial infarct), but not *possible infarto* (possible infarct)'.

Chute et al. (1998) propose a set of four main characteristics that medical terminology should comply with: general (utility and appropriateness in clinical application), structural, maintenance, and administration. A more detailed list of aspects is provided within each characteristic. On the general level, the terms are expected to be *complete* (with enough in-depth information and detail to make the term distinguishable), *comprehensive* (including all disciplines), *integrable* (supporting unique term representation), *non-redundant* (internal consistency in representation with supporting synonymy), and *mappable* (allowing data to be mapped to different systems for further use). The structural level suggests a model for easy-to-use interface that can be used for navigation, entry, and retrieval of terminology, taking into account the concepts of *atomic base, compositionality, synonymy, attributes, multiple hierarchies, explicit uncertainty, lexical rules,* and *representation.* The maintenance level provides guidance on how to keep terminology usable over time regardless of the changes that may occur within the discipline, while the administration level gives more detail on the coordination of interrelated terminologies, their access, and funding.

In this project, unlike the rest of the projects mentioned here, we did not rely on any previously prepared terminological database, although they do exist for the Croatian language (e.g., *Struna,* a database of Croatian Special Field Terminology, publicly available at: http://struna.ihjj.hr/). Rather, we wanted to learn directly from the corpus what types of multiword units occur in the text. This is the reason why we did not discriminate against any special set of nouns or adjectives. Still, since we

wanted to learn whether there are more nouns or adjectives from the medical domain in detected MWUs, we added those tags to the transducer's output for each
noun and adjective in the term. Similar to the process of deciding whether a multi-
word candidate term is from the medical domain, as performed for the Spanish
section of the MultiMedica corpus (Moreno Sandoval et al. 2018), we hope that this
information will do the same for Croatian MWUs.

5.6.2 Algorithm Design

The proposed algorithms are constructed as three main graphs in the NooJ environment, each of which presents a different MWU type:

1. NP followed by a Genitive NP, as in, e.g., (5.12)

$$< simptomatsko^{(A)} \, liječenje_{NP}^{(N \, NOM)} > <bolesnika_{NP\_GEN}^{(N \, GEN)}>$$
$$< \text{symptomatic} \quad \text{treatment} > <\text{of patients}> \tag{5.12}$$

2. NP followed by a PP, as in, e.g., (5.13)

$$< biljni^{(A)} \, sirup_{NP}^{(N \, NOM))} > <protiv^{(PREP)} kašlja_{PP}^{(N))}>$$
$$< \text{herbal} \quad \text{syrup} > <\text{against} \quad \text{cough}> \tag{5.13}$$

3. NP followed by a Genitive NP followed by one (or more) Genitive NPs, as
   in, e.g., (5.14)

$$<kontinuirano^{(A)} \, praćenje_{NP}^{(N \, NOM)} > <omjera_{NP\_GEN}^{(N \, GEN)} > <koristi_{NP\_GEN}^{(N \, GEN)} >$$
$$i^{(C)} < rizika_{NP\_GEN}^{(N \, GEN)} > <lijeka_{NP\_GEN}^{(N \, GEN)}>$$
$$<\text{continuous} \quad \text{evaluation}> <\text{of balance}> <\text{benefit}> \text{ and } <\text{risk }><\text{of a drug}>$$
$$\tag{5.14}$$

In each algorithm, an <NP> describes a noun preceded by one or up to four adjectives. An additional constraint is placed among the members of an NP: each adjective must match in case, gender, and number with the main noun of that NP in
order to be recognized as a single NP. The power of a transducer, characteristic of
syntactic grammars in NooJ, allows us to make additional annotations along each of
the paths. Thus, we are able to mark the path for each MWU, which in return provides
us with a list of specific tags for each adjective and a noun that builds a specific
MWU. This was very useful to our project, since we were interested to learn if an
adjective and/or a noun is from the medical domain (+med) or not (+general) and at
which position in an MWU it is found. It is important to note here that the closer the
adjective is to a noun, the lower position number it has (that is, in the NP 'pretty little
boy', the adjective 'little' is in position 1, and 'pretty' is in position 2).

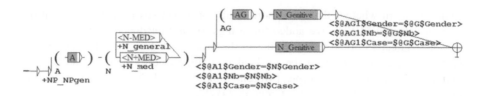

FIGURE 5.1   The main grammar of Algorithm 1, which recognizes an NP – NP_Genitive type of a multiword unit.

Figure 5.1 shows the main section of Algorithm 1. Nodes marked as A, N, and AG are variables that help us make additional constraints on the elements of a graph (in our case, the constraint is the match in gender ($Gender$), number (Nb), and case ($Case$) between a noun (N) and all the adjectives ($@A1$) that precede it). Figure 5.2 shows the subgraph that is responsible for recognizing the adjectives before a noun and annotating each as a general-domain adjective (+A_general) or medical-domain adjective (+A_med). The same graph is used for subgraph AG (inside the AG variable) with the only change being made in the variable names.

After the first NP is recognized, the graph can take either of the two available paths: (1) additional NP consisting of a single noun in genitive, or (2) a noun in the genitive case preceded again by any number from one to four adjectives (Figure 5.1). To make sure that we do not confuse the elements of each NP, adjectives from the second NP are saved into the variable $AG, while the noun is saved in the variable $G. The same constraints used in the first NP apply in the genitive NP as well. Multiword unit recognized with Algorithm 1 is tagged as <NP_NPgen>.

Sub-graphs A (Figure 5.2.) and AG are in essence the same grammar with the only change being in the name of the variable given to the final adjective (variables A1 and AG1, respectively).

Following the same logic inside an NP, Figure 5.3 presents a grammar that recognizes an NP followed by a prepositional phrase PP. The difference between

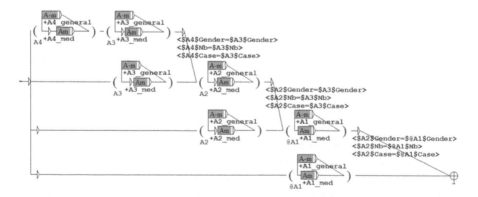

FIGURE 5.2   The sub-graph A, which recognizes multiple adjectives with the constraint on each to match in gender, number, and case.

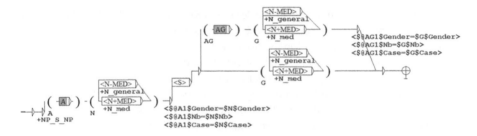

FIGURE 5.3   The main grammar of Algorithm 2, which recognizes an NP – PP type of multiword unit.

FIGURE 5.4   The main grammar of Algorithm 3, which recognizes an NP – NP_inGenitive1 – NP_inGenitive_Multiple type of multiword unit.

Algorithm 1 and Algorithm 2 is the additional node with a preposition <S> placed between the first NP and the second one. Multiword unit recognized with Algorithm 2 is tagged as <NP_S_NP>.

The third grammar (Figure 5.4) describes the Algorithm 3 that may start either with an NP or a PP (determined by the starting node <S>). Again, the same logic inside the adjective and noun nodes is respected. After recognizing the first NP (or PP), the grammar is looking for at least two more NP constructions in the genitive case, after which the grammar can take one of three possible endings: (1) it can exit, (2) it can find (any) number of additional NPs in the genitive before exiting, or (3) it can find one conjunction <C>, after which one or two more NPs in the genitive are found before exiting. Multiword unit recognized with Algorithm 3 is tagged as <NP_NPgen_multiple>.

The results per each algorithm type as well as their precision are discussed in the following section.

5.7   RESULTS

As already communicated, one of the objectives of this work is to enable the building of a medical lexicon that will be used for various purposes and NLP tasks. The creation of this lexicon is an ongoing project divided into several phases. In previous phases, the Croatian medical corpus was collected, and it is now continuously being made available through the Sketch Engine interface as the documents are tagged with the domain and subdomain markers. Medical terms consisting of a single noun, as well as those consisting of a noun and one or more adjectives, have been annotated in the corpus, extracted, and stored in the lexicon (Kocijan et al. 2020; 2021). At this time, we are expanding our focus to MWUs such as those in (5.12) and (5.13), as well as more complex ones like the one shown in (5.14).

Since there is no gold standard for concept boundaries in the medical domain for Croatian, we cannot compare our results to any such model. However, we can compare our precision results to other reports already mentioned in this chapter. Our results will be discussed with regard to true positives and false positives (since there is no manually tagged corpus, we are unable to provide results for true negatives and false negatives at this time). We will observe the results from the perspective of exact boundary matching – that is, if the recognized result completely matches the design of the grammar. Cases in which we have observed one-sided boundary matching (any one side of the algorithm may have an additional NP/PP), in the sense that we have recognized the shortcomings of the algorithm, which could be augmented with new nodes and/or paths, will be discussed through future steps sections that will be provided after each algorithm's results.

The number of recognized instances of MWUs by each algorithm is given in Table 5.2. Since it was almost double in size in comparison to other sub-sections we collected, sub-section M yielded the largest number of instances recognized by Algorithms 1 and 3, while the second-largest sub-section – S – yielded the largest number of instances recognized by Algorithm 2.

Except for the individual sub-section data provided in Table 5.2, the remaining results will be merged since all sub-sections are of the same type.

The distribution of the three categories of MWU we have discussed here is presented in Table 5.3. An MWU categorized as 'general' is one in which no adjectives

TABLE 5.2

Distribution of matching patterns in each subcorpus per algorithm

	A	H	M	S	V
Algorithm 1	561	675	**2 311**	894	515
Algorithm 2	390	109	568	**573**	376
Algorithm 3	61	39	**148**	117	84

TABLE 5.3

Distribution of three MWU types

Domain	<NP><NP+Gen>		<NP><PP>		<NP><NP+Gen><NP+Gen*	
	Total	Unique	Total	Unique	Total	Unique
General	798	317	426	204	21	7
Medical	4 158	1 960	1 590	985	428	88
TOTAL	4 956	2 277	2 016	1 189	449	95

or nouns from the medical domain are found. A medical MWU is a unit in which at least one of its parts is from the medical domain. Both the total number of detected units and the number of unique units are given for each algorithm. Due to the nature of the texts describing the medicament instructions, there are multiple occurrences of the same strings. Thus, we filtered out literal matches to see how many unique, i.e., different, items we have. However, the list still includes variations in case or number of the same construction (e.g., $kroničnoj^A bolesti^{N\_DAT} bubrega^{N\_GEN}$ and $kroničnu^A bolest^{N\_ACC} bubrega^{N\_GEN}$ are counted as different items on the unique list).

5.7.1 ALGORITHM 1

A more detailed analysis has provided us with the distribution of domain carriers within each category. There are 75 different patterns recognized by Algorithm 1 (Figures 5.1 and 5.2). Of these, only seven are found in the general-only domain (i.e., with none of the words belonging to the medical domain (see Table 5.6), while the remaining 68 patterns have at least one word that carries the additional semantic tag +MED (see Table 5.5).

It is interesting to note that we needed only one rule $[A^{General} N^{General} N^{Medical}]$ to recognize 30% of the terms (1 227), but also 29 rules (44% of the rules) to recognize only 1% of the terms in the medical domain. Similar findings were observed for the general population of terms, where only one rule $[A^{General} N^{General} N^{General}]$ was needed to describe 89% of the terms (711), while three rules (43%) described only 1% of the terms. The full distribution of patterns, given as the number and the percentage of a recognized pattern, are given in Table 5.4. The results have been categorized into six different ranges according to the number of occurrences for that pattern in the corpus. While medical-related patterns are found in all six ranges, the general ones are not as spread out and are found only in two, which was expected due to the nature of the text.

TABLE 5.4

Distribution of recognized patterns among medical and general terms detected by the Algorithm 1

No. of occurrences in the corpus	Patterns		Occurrences		Patterns		Occurrences	
	Medical terms				General terms			
	No.	%	No.	%	No.	%	No.	%
>1 000	1	1%	1 227	30%	0	0	0	0
500–1 000	1	1%	854	21%	1	14%	711	89%
100–500	5	7%	1 123	27%	0	0	0	0
50–100	9	13%	590	14%	0	0	0	0
5–50	23	34%	308	7%	3	43%	81	10%
1–4	29	43%	56	1%	3	43%	6	1%

TABLE 5.5

Distribution of the top ten medical and non-medical adjectives and nouns within the first category <NP><NP_Gen>, in which at least one element is from the medical domain

<NP>					<NP+Gen>						
A4	A3	A2	A1	N	A3	A2	A1	N_Gen	Total	A+med	N+med
1	**0**	**8**	**29**	**38**	**0**	**4**	**19**	**38**			
–	–	–	A^g	N^g	–	–	–	N^m	1 227	–	1 227
–	–	–	A^g	N^m	–	–	–	N^m	854	–	1 708
–	–	–	A^g	N^m	–	–	–	N^g	362	362	362
–	–	–	A^m	N^m	–	–	–	N^m	267	267	534
–	–	–	A^g	N^g	–	–	A^g	N^m	190	–	190
–	–	–	A^m	N^g	–	–	–	N^m	187	187	187
–	–	A^g	A^g	N^g	–	–	–	N^m	117	–	117
–	–	–	A^g	N^m	–	–	A^g	N^m	84	–	168
–	–	–	A^m	N^m	–	–	–	N^g	80	80	80
–	–	A^g	A^g	N^m	–	–	–	N^m	73	–	146

TABLE 5.6

Distribution of patterns within general-only class

<NP>					<NP+Gen>				
A4	A3	A2	A1	N	A3	A2	A1	N_Gen	
									822
-	-	-	A^g	N^g	-	-	-	N^g	711
-	-	A^g	A^g	N^g	-	-	-	N^g	58
-	-	-	A^g	N^g	-	-	A^g	N^g	38
-	A^g	A^g	A^g	N^g	-	-	-	N^g	9
-	-	A^g	A^g	N^g	-	-	A^g	N^g	4
-	A^g	A^g	A^g	N^g	-	-	A^g	N^g	1
-	-	-	A^g	N^g	-	A^g	A^g	N^g	1

Table 5.5 shows the top ten patterns that occurred in the text for the first category of MWUs, <NP> <NP+GEN>. The second row presents the position of the adjective and the noun, while the third row gives the total number of medical terms in each given position. Among the recognized patterns, in 38 of them the first noun is from the medical domain. The same is through for the second noun. There is only one adjective that is four places before the noun and it carries the medical marker. Among our patterns, there is not a single adjective with a medical marker that is

three places before the noun while the adjectives directly preceding the noun most frequently have a medical marker in an MWU. These results speak in favor of nouns as more dominant domain carriers in MWUs (within the medical domain) when compared to adjectives. Three final columns of Table 5.5 give the total number of recognized patterns and number of domain carriers within the adjective (A+med) and noun (N+med) group of words.

Some of the instances recognized by Algorithm 1, include examples shown in (5.15)–(5.18).

$$dugotrajna^{(A)} \, lokalna^{(A)} \, primjena^{(N \; NOM)} \, dorzolamida^{(N \; GEN)}$$
long–term local application of dorzolamide
$$(5.15)$$

$$generalizirana^{(A)} \, kalcifikacija^{(N \; NOM)} \, krvnih^{(A)} \, \check{z}ila^{(N \; GEN)}$$
generalized calcification of blood vessels
$$(5.16)$$

$$idiopatske^{(A)} \, prekomjerne^{(A)} \, aktivnosti^{(N \; NOM)} \, mokra\acute{c}nog^{(A)} \, mjehura^{(N \; GEN)}$$
idiopathic excessive activities of the bladder

$$(5.17)$$

$$ispravna^{(A)} \, primjena^{(NNOM)} \, odgovaraju\acute{c}e^{(A)} \, asepti\check{c}ke^{(A)} \, tehnike^{(NGEN)}$$
correct application of appropriate aseptic technique
$$(5.18)$$

A preliminary analysis of the recognized MWUs listed above reveals that all of them are good candidates to be considered as medical terms and thus to be included in the medical lexicon. However, prior to this step, the obtained results require manual analysis and classification. Some characteristic examples are discussed in the following section:

The MWU (5.15) consists of two NPs: (1. NP dugotrajna$^{(A)}$ lokalna$^{(A)}$ primjena$^{(N \; NOM)}$) and (2. NP dorzolamida$^{(N \; GEN)}$). When considering the structure of entries in the medical lexicon, the noun *primjena* (application) is a perfect candidate for a headword of the lexical entry. MWUs with modifying adjectives *dugotrajna* 'long-term' and *lokalna* 'local' can be stored as subentries of the headword. Subentries can be structured according to their complexity – e.g., (5.19).

$$primjena^{(N \; NOM)} - application$$
$$lokalna^{A)} primjena^{(N \; NOM)} - local \; application$$
$$dugotrajna^{A)} lokalna^{A)} primjena^{(N \; NOM)} - long - term \; local \; application.$$
$$(5.19)$$

On the other hand, the noun *dorzolamid* (dorzolamide) is also a perfect candidate for a headword in our medical lexicon. If we proceed with the analysis of the whole MWU *dugotrajna lokalna primjena dorzolamida* taking into account this fact, then the noun *dorzolamid* can be observed as a kind of semantic head of the whole expression, regardless of the fact that it is in an oblique case (genitive) and thus usually serves as a postmodification elsewhere. In this line of analysis, the entire NP with the noun in the nominative case preceding a noun in the genitive case can be understood as its modifier. In this case, the lexical entry would look like this:

$dorzolamid^{(N\ NOM)}-dorzolamide$
$\quad primjena^{(N\ NOM)}\ dorzolamida^{(N\ GEN)}-application\ of\ dorzolamide$
$\quad lokalna^{(A)}\ primjena^{(N\ NOM)}\ dorzolamida^{(N\ GEN)}-local\ application\ of\ dorzolamide$
$\quad dugotrajna^{(A)}\ lokalna^{(A)}\ primjena^{(N\ NOM)}\ dorzolamida^{(N\ GEN)}-long-term\ local$
$\qquad\qquad\qquad\qquad\qquad\qquad\qquad application\ of\ dorzolamide,\ etc.$

$$(5.20)$$

At this stage in the development of the lexicon, we opt for both solutions, i.e., we have decided to introduce two separate entries for two headwords and to store parts of complex MWUs as their subentries, as presented above. The advantage of such a procedure is that such complex MWUs are stored at least once as a single unit in the lexicon and therefore can be used later for various purposes, such as machine translation.

This solution is not applied to MWUs such as *kronični glaukom otvorenog kuta* – chronic open angle glaucoma. This MWU consists of two noun phrases (1. $(kronični^{(A)}\ glaukom^{(N\ NOM)}{}_{NP})$ + 2. $(otvorenog^{(A)}\ kuta^{(N\ GEN)}{}_{NP})$. The NP in the oblique case is a postmodification of the preceding NP in the nominative and thus is not stored separately. Therefore, there is only one headword (*glaukom* – *glaucoma*) provided in the lexicon. Its lexical entry looks like in example (5.21).

$glaukom^{(N\ NOM)}-$ glaucoma
$\quad kronični^{(A)}\ glaukom^{(N\ NOM)}-$ chronic glaucoma
$\quad kronični^{(A)}\ glaukom^{(N\ NOM)}\ otvorenog^{(A)}\ kuta^{(N\ GEN)}-$ chronic open – angle glaucoma
$\quad hipertenzivni^{(A)}\ primarni^{(A)}\ glaukom^{(N\ NOM)}\ otvorenoga^{(A)}\ kuta^{(N\ GEN)}-$ high – pressure
$\qquad\qquad\qquad\qquad\qquad\qquad\qquad\qquad primary\ open-angle\ glaucoma\ etc.$

$$(5.21)$$

When dealing with NPs where only adjectives belong to the medical domain, whereas nouns are not recognized as medical terms – i.e., they do not have a +MED tag in the dictionary, the whole NP is taken as a lexical entry. This pertains to noun phrases as *dišni put* – 'airway', *usna šupljina* – 'oral cavity', etc. This also pertains to cases when postmodifiers are tagged as +MED (e.g., *šupljina grkljana* – 'laryngeal cavity'). At this stage of development, the lexical entries for NPs containing the noun *šupljina* 'cavity' look like this:

$usna^{(A)}\ šupljina^{(N\ NOM}-$ mouth cavity
$\quad šupljina^{(N\ NOM)}\ prsnog^{(A)}\ koša^{(NGEN)}-$ thoracic cavity $\qquad\qquad$ (5.22)
$\quad šupljina^{(N\ NOM)}\ grkljana^{(N\ GEN)}-$ laryngeal cavity etc.

In Table 5.6 the distribution of patterns with only general vocabulary is provided (i.e., neither adjectives nor nouns had been previously tagged as +MED).

5.7.2 ALGORITHM 2

In this section, we will provide the same distributional analysis for Algorithm 2.

Although this algorithm did not find as many occurrences as Algorithm 1, the number of patterns was greater. There are 81 different patterns that were recognized by Algorithm 2. Of those, eight are found in the general-only domain (Table 5.9), while the remaining patterns have at least one word that carries the additional semantic tag +MED (Table 5.8). Again, the number of rules is inversely proportional to the number of recognized terms in both medical and general domains. The full distribution of patterns and the percentages of recognized patterns is given in Table 5.7. Since the largest number of recognized patterns was 342 (for the general-only domain), Table 5.7 includes only data for the categories lower than 500.

Table 5.8 shows the top ten performing patterns where at least one adjective or noun has a medical domain marker. Again, the number of medical markers in each position is given for the full list of 81 patterns.

The algorithm recognizes instances like those in (5.23)–(5.25):

$$abnormalan^{(A)} \; osje\acute{c}aj^{(N\;NOM)} \; u^{(PREP)} \; oku^{(N\;LOC)} - \text{abnormal sensation in the eye}$$

$$(5.23)$$

$$akutna^{(A)} \; bol^{(N\;NOM)} \; poslije^{(PREP)} \; stomatolo\check{s}kog^{(A)} \; kirur\check{s}kog^{(A)} \; zahvata^{(N\;GEN)}$$
acute pain after a stomatological surgical procedure

$$(5.24)$$

TABLE 5.7

Distribution of recognized patterns among medical and general terms detected by Algorithm 2

No. of occurrences in the corpus	Patterns		Occurrences		Patterns		Occurrences	
	Medical terms				General terms			
	No.	%	No.	%	No.	%	No.	%
100–500	4	5%	775	49%	1	13%	342	80%
50–100	4	5%	272	17%	0	0	0	0
5–50	25	31%	443	28%	2	25%	79	19%
1–4	48	59%	100	6%	5	63%	7	2%

TABLE 5.8

Top ten medical vs non-medical adjectives and nouns within the second category <NP><PP> where at least one element is from the medical domain

<NP>									<PP>			
A4	A3	A2	A1	N	S	A3	A2	A1	N_Gen		A+med	N+med
0	**1**	**15**	**40**	**51**	**0**		**3**	**26**	**42**			
–	–	–	A^g	N^g	–	–	–		N^m	274	–	–
–	–	–	A^g	N^m	–	–	–		N^m	247	–	494
–	–	–	A^g	N^m	–	–	–		N^g	140	–	140
–	–	–	A^m	N^m	–	–	–		N^m	114	114	228
–	–	–	A^g	N^m	–	–		A^g	N^g	94	–	94
–	–	–	A^m	N^m	–	–	–		N^g	75	75	75
–	–	–	A^m	N^g	–	–	–		N^m	53	53	53
–	–	–	A^g	N^m	–	–		A^g	N^m	50	–	100
–	–	–	A^m	N^g	–	–	–		N^g	45	45	–
–	–	–	A^g	N^g	–	–		A^g	N^m	41	–	41

TABLE 5.9

Distribution of patterns within general-only class of <NP><PP>

<NP>					<PP>			Total no.
A3	A2	A1	N	S	A2	A1	N	426
–	–	A^g	N^g		–	–	N^g	342
–	–	A^g	N^g		–	A^g	N^g	48
–	A^g	A^g	N^g		–	–	N^g	29
–	–	A^g	N^g		A^g	A^g	N^g	2
–	A^g	A^g	N^g		–	A^g	N^g	2
–	A^g	A^g	N^g		A^g	A^g	N^g	1
A^g	A^g	A^g	N^g		–	A^g	N^g	1
A^g	A^g	A^g	N^g		–	–	N^g	1

$$apsolutni^{(A)}\ rizik^{(N\ NOM)}\ od^{(PREP)}\ kardiovaskularnih^{(A)}\ malformacija^{(N\ GEN)}$$
apsolute risk of cardiovascular malformations

$$(5.25)$$

These and similar MWUs are processed and stored in the medical lexicon according to the principles described above. The general structure of these MWUs is: noun phrase (nominative) + prepositional phrase (preposition + noun (oblique)).

Although prepositional phrases function as postmodifications of preceding noun phrases in the nominative case, oblique NPs within prepositional phrases are also excellent candidates for medical terms. Therefore, we decided to analyze them separately and store them in the lexicon accordingly. This means that the MWU given in (5.23) serves as the basis for two headwords in the lexicon (*osjećaj* 'sensation' and *oko* 'eye'). The whole MWUs are recorded as subentries of NPs in the nominative case, as in (5.26).

$osjećaj^{(N\ NOM)}$ – sensation

   $abnormalan^{(A)}\ osjećaj^{(NNOM)}$ – abnormal sensation

   $abnormalan^{(A)}\ osjećaj^{(N\ NOM)}\ u^{(PREP)}\ oku^{(N\ LOC)}$ – abnormal sensation in the eye

(5.26)

The MWU given in (5.24) is presented in the lexicon as shown in (5.27) and (5.28).

$bol^{(N\ NOM)}$ – *pain*

   $akutna^{(A)}\ bol^{(N\ NOM)}$ – acute pain

   $akutna^{(A)}\ bol^{(N\ NOM)}\ poslije^{(PREP)}\ stomatološkog^{(A)}\ kirurškog^{(A)}$ (5.27)

      $zahvata^{(N\ GEN)}$ –

acute pain after the stomatological surgical procedure

$zahvat^{(N\ NOM)}$ – procedure

   $kirurški^{(A)}\ zahvat^{(NNOM)}$ – surgical procedure

   $stomatološki^{(A)}\ kirurški^{(A)}\ zahvat^{(N\ NOM)}$ – stomatological surgical procedure

(5.28)

Thus, although such complex multiword units are separated into two lexicon entries, their full structures are preserved at least once in the lexicon and can be used later if necessary.

Further, the analysis of the context before and after the recognized string shows room for enhancement. In some cases, the final NP can be followed by additional NPs in the genitive or a PP, or the first NP can be preceded by an NP. In some cases, both pre- and post-additions are possible, as in examples (5.29) – (5.33) (unrecognized strings are in parentheses).

$adrenalna^{(A)}\ insuficijencija^{(N\ NOM)}\ nakon^{(PREP)}\ prekida^{(N\ GEN)}$

$(liječenja^{(N\ GEN)}\ \mathbf{kortikosteroidima}^{(N\ INST)})$ (5.29)

adrenal insufficiency after corticosteroid treatment interruption

$aktivni^{(A)}$ $oblik^{(N\ NOM)}$ $u^{(PREP)}$ $kiselim^{(A)}$ $uvjetima$

($\mathbf{u}^{(PREP)}$ $\mathbf{parijetalnim}^{(A)}$ $\mathbf{stanicama}^{(N\ LOC)}$)   (5.30)

active form in acid conditions in parietal cells

$izložene^{(A)}$ $trudnoće$ $u^{(PREP)}$ $žena$ ($s^{(PREP)}$ $\mathbf{komplikacijama}^{(N\ INST)}$

$\mathbf{krvarenja}^{(N\ GEN)}$)   (5.31)

exposed pregnancies of women with bleeding disorders

$apsolutni^{(A)}$ $rizik^{(N\ NOM)}$ $od^{(PREP)}$ $kardiovaskularnih^{(A)}$ $malformacija^{(N\ GEN)}$

absolute r of cardiovascular malformations

(5.32)

($\mathbf{stabilizacija}^{(N\ NOM)}$) $astmatičnih^{(A)}$ $bolesnika^{(N\ GEN)}$ $prije^{(PREP)}$ $primjene^{(N\ GEN)}$

($\mathbf{kontrastnih}^{(A)}$ $\mathbf{sredstava}^{(N\ GEN)}$)

stabilization of asthmatic patients before the application of contrast agents

(5.33)

We have also observed different types of coordination that we will need to incorporate in the algorithm in some future steps. In some cases, more adjectives are involved, as in (5.34) and (5.35) (unrecognized strings are in parentheses).

($\boldsymbol{bistre}^{(A)}$ $\boldsymbol{i}^{(CONJ)}$) $bezbojne^{(A)}$ $kapi^{(N\ NOM)}$ $za^{(PREP)}$ $oko^{(N\ ACC)}$

clear and colorless eye drops   (5.34)

($\boldsymbol{bistra}^{(A)}$,) $bezbojna^{(A)}$ $vodena^{(A)}$ $otopina^{(N\ NOM)}$ $bez^{(PREP)}$ $vidljivih^{(A)}$ $čestic$

$a^{(N\ GEN)}$

clear, colorless water solution without visible residues

(5.35)

but also one or more NPs and/or PPs, as in examples (5.36)–(5.38) (unrecognized strings are in parentheses).

$želučani^{(A)}$ $ulkusi^{(N\ NOM)}$ $s^{(PREP)}$ $perforacijom^{(NINST)}$ ($\boldsymbol{i}^{(CONJ)}$ $\boldsymbol{krvarenjem}^{(N\ INST)}$)

gastric ulcers with perforation (and bleeding)

(5.36)

$štetni^{(A)}$ $utjecaj^{(N\ NOM)}$ $na^{(PREP)}$ $trudnoću^{(N\ ACCU)}$ ($\boldsymbol{i}^{(CONJ)}$

$/\boldsymbol{ili}^{(CONJ)}$ $\mathbf{embriofetalni}^{(A)}$ $\mathbf{razvoj}^{(N\ ACCU)}$)   (5.37)

adverse impact to pregnancy (and/or embryo − fetal development) (**plućni**$^{(A)}$ **edem**$^{(N\ NOM)}$, **hemodinamske**$^{(A)}$ **promjene**$^{(N\ NOM)}$,) *ishemijske*$^{(A)}$

promjene$^{(N\ NOM)}$ *na*$^{(PREP)}$ *EKG − u*$^{(N\ LOC)}$ (**te**$^{(CONJ)}$ **različite**$^{(A)}$ **aritmije**$^{(N\ NOM)}$)

(pulmonary edema, hemodynamic changes,) ischemic changes in ECG

(and various arrhythmias)

(5.38)

As the preliminary analysis shows, all recognized strings provide an excellent source for populating the lexicon. Expansion and adjustment of algorithms for their recognition as well as their thorough analysis are beyond the scope of this paper. In Table 5.9, below, we present the distribution of recognized MWUs whose members were not previously tagged as belonging to the medical domain.

5.7.3 ALGORITHM 3

In this section, we will provide the same distributional analysis for Algorithm 3.

As might be expected, this algorithm found the least number of occurrences (compared to Algorithms 1 and 2) but the highest number of patterns, i.e., 95 in total, 88 of which were from the medical domain (Table 5.11) and seven from the general-only domain (Table 5.12). Again, the number of rules is inversely proportional to the number of recognized terms in both medical and general domains. The full distribution of patterns and the percentages of recognized patterns is given in Table 5.10. Since the largest number of recognized patterns was 81 (for the medical domain), Table 5.10 includes only data for the categories lower than 100.

Table 5.11 shows the top ten performing patterns where at least one adjective or a noun has a medical domain marker. Again, the number of medical markers in each position is given for the full list of 88 patterns.

Some of the recognized instances are presented through (5.39)–(5.43).

TABLE 5.10

Distribution of recognized patterns among medical and general terms detected by Algorithm 3

No. of occurrences in the corpus	Patterns		Occurrences		Patterns		Occurrences	
	Medical terms				*General terms*			
	No.	%	No.	%	No.	%	No.	%
50–100	2	2%	154	36%	0	0	0	0
5–50	12	14%	156	36%	2	29%	12	57%
1–4	74	84%	118	28%	5	71%	9	43%

TABLE 5.11

Distribution of medical vs non-medical adjectives and nouns inside the third category <NP><NP+Gen><NP+Gen>*

A3	A2	A1	N	A2	A1	N_Gen1	A2	A1	N_Gen2	A1	N_Gen3	A2	A1	N_Gen4	N_med	A+med	N+med
0	4	27	47	0	7	41	0	17	62	2	19	0	1	6			
–	–	A^g	N^m	–	–	N^g	–	–	N^m	–	–	–	–	–	81	–	162
–	–	A^g	N^g	–	–	N^g	–	–	N^m	–	–	–	–	–	73	–	73
–	–	A^g	N^m	–	–	N^g	–	–	N^g	–	N^g	–	–	–	39	–	39
–	–	A^g	N^g	–	A^g	N^m	–	–	N^m	–	–	–	–	–	27	–	54
–	–	A^g	N^m	–	–	N^g	–	–	N^m	–	–	–	–	–	23	–	46
–	A^g	A^m	N^m	–	A^g	N^m	–	–	N^m	–	–	–	–	–	12	12	36
–	–	A^g	N^g	–	–	N^m	–	–	N^m	–	–	–	–	–	10	–	20
–	–	A^g	N^m	–	–	N^m	–	–	N^m	–	–	–	–	–	9	–	27
–	–	A^m	N^m	–	–	N^g	–	–	N^g	–	–	–	–	–	9	9	18
–	–	A^g	N^g	–	–	N^m	–	–	N^g	–	–	–	–	–	7	–	7

TABLE 5.12
Distribution of patterns within the general-only class for the third algorithm

<NP> <NP+Gen><NP+Gen><NP+Gen><NP+Gen>

A2	A1	N	Ngen1	Ngen2	Ngen3	Ngen4	No. of occurances
							21
–	A^g	N^g	N^g	N^g	–	–	7
–	A^g	N^g	N^g	N^g	N^g	–	5
–	A^g	N^g	$A^g\ N^g$	N^g	–	–	3
A^g	A^g	N^g	N^g	N^g	N^g	N^g	2
–	A^g	N^g	N^g	$A^g\ N^g$	–	–	2
–	A^g	N^g	N^g	N^g	N^g	–	1
A^g	A^g	N^g	N^g	N^g	–	–	1

$$\textit{pojačano}^{(A)}\ \textit{oticanje}^{(N\ NOM)}\ \textit{sluznice}^{(N\ GEN)}\ \textit{oka}^{(N\ GEN)}$$
$$\text{increased swelling of mucous eye membrane} \qquad (5.39)$$

$$\textit{pojedinačni}^{(A)}\ \textit{slučajevi}^{(N\ NOM)}\ \textit{simptoma}^{(N\ GEN)}\ \textit{aseptičkog}^{(A)}\ \textit{meningitisa}^{(N\ GEN)}$$
$$\text{individual cases of aseptic meningitis symptoms}$$
$$(5.40)$$

$$\textit{lokalna}^{(A)}\ \textit{nadraženost}^{(N\ NOM)}\ \textit{gornjeg}^{(A)}\ \textit{dijela}^{(N\ GEN)}\ \textit{gastrointestinalne}^{(A)}\ \textit{sluznice}^{(N\ GEN)}$$
$$\text{local irritation of upper gastrointestinal mucosa}$$
$$(5.41)$$

$$\textit{gljivične}^{(A)}\ \textit{infekcije}^{(N\ NOM)}\ \textit{usne}^{(A)}\ \textit{šupljine}^{(N\ GEN)}\ i^{(CONJ)}\ \textit{probavnih}^{(A)}\ \textit{organa}^{(N\ GEN)}$$
$$\text{fungal infections of mouth cavity and digestive organs}$$
$$(5.42)$$

$$\textit{teški}^{(A)}\ \textit{poremećaj}^{(N\ NOM)}\ \textit{funkcije}^{(N\ GEN)}\ \textit{mokraćnog}^{(A)}\ \textit{mjehura}^{(N\ GEN)}\ i^{(CONJ)}$$
$$\textit{opstrukcija}^{(N\ NOM)}\ \textit{mokraćne}^{(A)}\ \textit{cijevi}^{(N\ GEN)} \qquad (5.43)$$
$$\text{severe disorder of bladder function and urethra obstruction}$$

We have noticed that the algorithm should be enhanced with an additional noun at the end of the string, and also an additional prepositional phrase at the beginning or ending, like in the instances shown in (5.44) and (5.45) (unrecognized strings are in parentheses).

$$(\boldsymbol{\textit{bolesnici}}^{(N\ NOM)})\ s^{(PREP)}\ \textit{rijetkim}^{(A)}\ \textit{nasljednim}^{(A)}\ \textit{poremećajem}^{(N\ INST)}$$
$$\textit{nepodnošenja}^{(N\ GEN)}\ \textit{fruktoze}^{(N\ GEN)} \qquad (5.44)$$
$$\text{(patients) with rare hereditary condition of fructose intolerance}$$

$ozbiljnije^{(A)} posljedice^{(N\ NOM)} čestih^{(A)} nuspojava^{(N\ GEN)} kortikosteroida^{(N\ GEN)}$

$(\boldsymbol{u}^{(PREP)} \boldsymbol{starijoj}^{(A)} \boldsymbol{dobi}^{(N\ LOC)})$ (5.45)

more serious consequences of frequent side effects of corticosteroids (in elderly age)

Since our corpus was not manually checked, we do not know how many results we should recognize and are thus unable to provide recall at this time. However, after a preliminary correction (as explained in the section on results) the precision of the system is 100% for the first two algorithms, and 99.785% for the third one, which is better precision-wise than all other rule-base systems' reports. The highest F-score of 82.1% for a rule-based system was designed by Doan, Bastarache, Klimkowski, Denny, and Xu in 2009, and their team's score was ranked second at 2009 i2b2 NLP challenge (Informatics for Integrating Biology and the Bedside = i2b2). To our knowledge, no other reports have been made with better scores so far.

5.7.4 ERROR ANALYSIS

After the first run of the proposed models, we detected errors that were due to the conversion of a pdf file. Tables, headings, and bullets were converted as running text that was missing punctuation marks. To help reduce these types of errors, we added a line break after each such occurrence in order to clearly mark the sections. This raised the precision in all three algorithms.

The errors detected in the third algorithm were due to the use of quotation marks around an adjective, which disabled the algorithm from finding the remaining of the term, as shown in (5.46).

$općenita^{(A)} preferenca^{(N\ NOM)} medijana^{(N\ GEN)} broja^{(N\ GEN)} \underline{\textbf{\textit{„zaslijepljenih"}}}^{(A)}$

$osoba^{(N\ GEN)}$

general median preference of "blinded" persons rate   (5.46)

This type of an error, however, will remain until the next phase of the project.

5.8 CONCLUSION

In this chapter, we presented a rule-based method for the detection of multi-word units belonging to the medical domain of the Croatian language. Our method for the detection of MWUs is based on morpho-syntactic rules developed in the NooJ platform in the form of finite-state transducers. The whole procedure consists of three algorithms, each of them aiming at recognizing different types of complex MWUs. The MWUs that we deal with are complex because they consist of NPs followed by one or more NPs or PPs. The results of applied algorithms are manually checked and further analyzed. The high precision scores for all three models allow us to use the results for the building of the medical lexicon and can potentially be used in various other NLP tasks in the medical domain.

We have demonstrated how different linguistic patterns can be used to build rule-based algorithms to automatically find the boundaries of multiword units in the medical domain, obtaining a 100% precision score. We briefly present the structure of entries in the medical lexicon for Croatian and furthermore argue that complex MWUs can be decomposed in order to make the structure of the medical lexicon more dense and coherent. At the same time, we argue that complex MWUs (consisting of multiple NPs and/or PPs) should be preserved and recorded as such, since they will be used in further phases of this ongoing project, and we believe that the length of an MWU is project-dependent. We have also indicated how the rules we have developed and applied here can be adapted and extended in order to tackle coordination within NPs and PPs and other phenomena. We plan to deal with the adjustment and expansion of rules in our future work.

NOTE

1 The abbreviations we use here refer to: ATTR – attribute, PL – plural, DIM – diminutive.

REFERENCES

Agić, Ž. and B. Bekavac. 2013. Domain-aware evaluation of named entity recognition systems for Croatian. *CIT. Journal of Computing and Information Technology* 21, no. 3: 195–209. doi:10.2498/cit.1002190

Almic, P., & Snajder, J. (2014). Determining the Semantic Compositionality of Croatian Multi-Word Expressions. In: Proceedings of the Ninth Language Technologies Conference, Information Society (IS-JT 2014), Ljubljana 32–37.

Alphonse, Y. and P. Bouillon. 2004. Methodology for building thematic indexes in medicine for French. In *LREC 2004 Fourth International Conference on Language Resources and Evaluation*, eds. M. T. Lino, M. F. Xavier, F. Ferreira, R. Costa and R. Silva, 789–792. ELRA - European Language Ressources Association. https://archive-ouverte.unige.ch/unige:2266 (accessed December 15, 2020).

Barčot, B. and T. Milčić. 2019. Arijadnina nit u ovladavanju frazemima na nastavi materinskoga i stranoga jezika - kvantitativno obrađeni podaci (prvi dio studije). *Filološke studije* XVII, no. 1: 292–304. doi:10.17072/1857-6060-2019-17-1-292-304

Barrows Jr, R. C., M. Busuioc and C. Friedman. 2000. Limited parsing of notational text visit notes: ad-hoc v-s. NLP approaches. In *Proceedings AMIA Symposium*, 51–55. https://www.researchgate.net/publication/12247608_Limited_parsing_of_notational_text_visit_notes_Ad-hoc_vs_NLP_approaches (accessed December 10, 2020).

Bekavac, B. and M. Tadić. 2009. A generic method for multi word extraction from wikipedia. In *Technologies for the Processing and Retrieval of Semi-Structured Documents: Experience from the CADIAL Project*, eds. M. Tadić, B. Dalbelo Bašić and M. Moens, 115–124. Zagreb: Croatian Language Technologies Society.

Borchert, F., C. Lohr, L. Modersohn, T. Langer, M. Follmann, J. P. Sachs, U. Hahn and M.-P. Schapranow. 2020. GGPONC: A corpus of german medical text with rich metadata based on clinical practice guidelines. In *Proceedings of the 11th International Workshop on Health Text Mining and Information Analysis*, 38–48, Association for Computational Linguistics. https://www.aclweb.org/anthology/2020.louhi-1.5.pdf (accessed December 10, 2020).

Borucinsky, M., and Kegalj, J. (2017). Viserjecni nazivi u jeziku brodostrojarske struke (Multiword lexical unitsinMarine Engin eering Texts). In Od teorije do prakse u jeziku struke, Omrcen, Darija, and Krakic, Ana-Marija (eds.). Zagreb: Udruga nastavnika jezika struke, 2017, pp. 7–23.

Boujelben, I., S. Mesfar and A. Ben Hamadou. 2011. Methodological approach of terminological extraction applied to biomedical domain. In *Proceedings of the 4th International Conference on Information Systems and Economic Intelligence* SIIE2011, Marrakech.

di Buono, M. P., A. Maisto and S. Pelosi. 2015. From linguistic resources to medical entity recognition: A supervised morpho-syntactic approach. In (eds.) Proceedings of the ALLDATA2015: The First International Conference on Big Data, Small Da ta, Linked Data and Open Data, ALLDATA 82: 81–86, IARIA. ThinkMind ISBN: 978-1-61208-445-9. Grzymala-Busse, J. Schwab, I. di Buono, M. P.

Chute, C. G., S. P. Cohn and J. R. Campbell. 1998. A framework for comprehensive health terminology systems in the United States: Development guidelines, criteria for selection, and public policy implications. ANSI Healthcare Informatics Standards Board Vocabulary Working Group and the Computer-Based Patient Records Institute Working Group on Codes and Structures. *Journal of the American Medical Informatics Association* 5: 503–510.

Doan, S., L. Bastarache, S. Klimkowski, J. C. Denny and H. Xu. 2010. Integrating existing natural language processing tools for medication extraction from discharge summaries. In *Journal of the American Medical Informatics Association* 17: 528–531.

Doan, S. and H. Xu. 2010. Recognizing medication related entities in hospital discharge summaries using support vector machine. In: *Coling 2010: Poster Volume*, 259–266.

Ehmann, B. and V. Garami. 2010. Narrative psychological content analysis with NooJ: Linguistic markers of time experience in self reports. In *Applications of Finite-State Language Processing: Selected Papers from the 2008 International NooJ Conference*, eds. T. Varadi, J. Kuti and M. Silberztein, 180–190, Newcastle: Cambridge Scholars Publishing.

El-Rab, W. G., O. R. Zaïane and M. El-Hajj. 2017. Formalizing clinical practice guideline for clinical decision support systems. *Health Informatics Journal* 23, no. 2, 146–156. 10.1177/1460458216632272 (accessed December 7, 2020).

Filipović Petrović, I. and J. Parizoska. 2019a. Konceptualna organizacija frazeoloških rječnika u e-leksikografiji. In *Filologija: časopis Razreda za filološke znanosti*, Zagreb: Hrvatske akademije znanosti i umjetnosti, 73, 27–45.

Filipović Petrović, I. and J. Parizoska. 2019b. Mogućnosti leksikografske obrade promjenjivih frazema u mrežnome frazeološkom rječniku hrvatskoga jezika. In *E-rječnici i e-leksikografija*. http://ihjj.hr/mreznik/uploads/7b5602698002896a0fd4f2dcee210796.pdf

Frančić, A. and M. Menac-Mihalić. 2013. Istraživanje i leksikografska obradba hrvatske dijalektne frazeologije. In *A tko to ide? A xto tam idze?* Zagreb: Hrvatska sveučilišna naklada; Hrvatsko filološko društvo, 67–84.

Friedman, C. 2005. Semantic text parsing for patient records. In: *Medical Informatics*, eds. H. Chen, S. S. Fuller, C. Friedman and W. Hersh. Integrated Series in Information Systems, vol 8. Boston, MA: Springer. 10.1007/0-387-25739-X_15 (accessed December 15, 2020).

Friedman, C. and G. Hripcsak. 1998. Evaluating natural language processors in the clinical domain. *Methods of Information in Medicine* 37, no. 4-5: 334–344.

Friedman, C., G. Hripcsak and I. Shablinsky. 1998. An evaluation of natural language processing methodologies. In: *Proceedings of AMIA Symposium*, 855–859. https://www.ncbi.nlm.nih.gov/pmc/articles/PMC2232366/

Friedman, C., T. C. Rindflesch and M. Corn. 2013. Natural language processing: State of the art and prospects for significant progress, a workshop sponsored by the National Library of Medicine. Special Communication in *Journal of Biomedical Informatics* 46: 765–773.

Genevieve, G., R. Angus, J. Richard and S. Robert. 2013. Finding negative symptoms of schizophrenia in patient records. In *Proceedings of the Workshop on NLP for Medicine and Biology associated with RANLP 2013*, pages 9–17. Hissar, Bulgari. INCOMA Ltd. Shoumen, BULGARIA. https://aclanthology.org/W13-5102

Haug, P. J., Ranum, D. L., & Frederick, P. R. 1990. Computerized extraction of coded findings from free-text radiologic reports. Work in progress. *Radiology*, 174, no. 2: 543–548. https://doi.org/10.1148/radiology.174.2.2404321

Imaichi, O., T. Yanase and Y. Niwa. 2013. A Comparison of Rule-Based and Machine Learning Methods for Medical Information Extraction. In: The First Workshop on Natural Language Processing for Medical and Healthcare Fields, 38–42. Nagoya. Asian Federation of Natural Language Processing. https://aclanthology.org/W13-4607

Jelčić, J. 2014. Croatian EFL learners' comprehension of idiom use: Context, decomposability and age factors. *Jezikoslovlje* 15, no. 2-3: 373–393.

Johnson II D. B., R. Taira, A. F. Cardenas and D. Aberle. 1997. Extracting information from free text radiology reports. *International Journal of Digital Libraries* 1: 297–308, Springer-Verlag.

Kang, N., B. Singh, Z. Afzal, E. M. v. Mulligen and J. A. Kors. 2013. Using rule-based natural language processing to improve disease normalization in biomedical text. *Journal of the American Medical Informatics Association* 20: 876–881.

Kocijan, K., & Librenjak, S. (2018). The quest for croatian idioms as multiword units. In *Multiword Units in Machine Translation and Translation Technology*. Ruslan, Mitkov, Monti, Johanna, Gloria, Corpas Pastor, and Seretan, Violeta (eds.) Amsterdam: John Benjamins Publishing Company, pp. 202–221. doi : 10.1075/cilt.341.10koc.

Kocijan, K., S. Kurolt and L. Mijić. 2020. Building Croatian medical dictionary from medical corpus. In *Rasprave: Časopis Instituta za hrvatski jezik i jezikoslovlje* 46/2, 765–782. 10.31724/rihjj.46.2.17 (accessed December 1, 2020).

Kocijan, K. and S. Librenjak. 2016. Recognizing verb-based Croatian idiomatic MWUs. In *Automatic Processing of Natural Language Electronic Texts with NooJ*. Switzerland, Springer, str. 96–106.

Kocijan, K., K. Šojat and S. Kurolt. 2021. Multiword expressions in the medical domain: Who carries the domain specific meaning, In *Formalizing Natural Languages: Applications to Natural Language Processing and Digital Humanities*, eds. B. Bekavac, K. Kocijan, M. Silberztein and K. Šojat, 49–60, Switzerland: Springer, Cham.

Koleck, T. A., C. Dreisbach, P. E. Bourne and S. Bakken. 2019. Natural language processing of symptoms documented in free-text narratives of electronic health records: A systematic review. *Journal of the American Medical Informatics Association*: JAMIA 26, no. 4: 364–379. 10.1093/jamia/ocy173 (accessed December 5, 2020).

Kovačević, B. and E. Ramadanović. 2013. Frazemske polusloženice (od rječnika preko tvorbe do pravopisa i obratno). *Rasprave Instituta za hrvatski jezik i jezikoslovlje* 39, no. 1: 369–389.

Lamare, J. B., T. Olatunji and L. Yao. 2020. On the diminishing return of labeling clinical reports. In *Proceedings of the 3rd Clinical Natural Language Processing Workshop*, 280–290, Online. Association for Computational Linguistics. DOI: 10.18653/v1/202 0.clinicalnlp-1.31

Leaman, R., L. Wojtulewich, R. Sullivan, A. Skariah, J. Yang and G. Gonzalez. 2010. Towards internet-age pharmacovigilance: Extracting adverse drug reactions from user posts to health-related social networks. In *Proceedings of the 2010 Workshop on Biomedical Natural Language Processing*, 117–125, ACL 2010, Association for Computational Linguistics. https://www.aclweb.org/anthology/W10-1915.pdf (accessed December 7, 2020).

Ljubešić, N., K. Dobrovoljc, S. Krek, M. Peršurić Antonić and D. Fišer. 2014. hrMWELex – A MWE lexicon of Croatian extracted from a parsed gigacorpus. In *Language Technologies: Proceedings of the 17th International Multiconference Information Society IS2014*. Institut 'Jožef Stefan'. Ljubljana, Slovenia, 25–31.

Marković, I. 2012. Uvod u jezičnu morfologiju, Disput, Zagreb.

Meystre, S. and P. J. Haug. 2006. Natural language processing to extract medical problems from electronic clinical documents: Performance evaluation. *Journal of Biomedical Informatics* 39: 589–599.

Mondal, A., D. Das and S. Bandyopadhyay. 2016. Relationship extraction based on category of medical concepts from lexical contexts. In *Proceedings of the 14th Intl. Conference on Natural Language Processing* (ICON-2017), 212–219, Kolkata, India: NLP Association of India. https://aclanthology.org/W17-7527

Parizoska, J. 2018. What constitutes the core of a verbal idiom? Stability and variation in a cross-linguistic perspective. In *EUROPHRAS 2018 "Reproducibility from a Phraseological Perspective: Structural, Functional and Cultural Aspects"*. Uniwersytet w Białymstoku Białystok, Poland.

Pilan, I., P. H. Brekke, F. A. Dahl, T. Gundersen, H. Husby, Ø. Nytrø and L. Øvrelid. 2020. Classification of syncope cases in Norwegian Medical Records. In *Proceedings of the 3rd Clinical Natural Language Processing Workshop*, 79–84, Association for Computational Linguistics. (accessed December 10, 2020) https://www.aclweb.org/anthology/2020.clinicalnlp-1.9.pdf

Radek, I. and I. Pešut. 2016. Kolokacije u engleskom jeziku policijske struke. *Policija i sigurnost* 25: 370–396.

Rajh, I. 2015. Looking for multiword terms in a comparable bilingual corpus. In *Computerised and Corpus-based Approaches to Phraseology: Monolingual and Multilingual Perspectives*. Europhras 2015, Malaga, Spain.

Sandoval, M., J. Díaz, L. Campillos Llanos and T. Redondo. 2018. Biomedical term extraction: NLP techniques in computational medicine. *International Journal of Interactive Multimedia and Artifcial Intelligence*, Special Issue on Artificial Intelligence Applications (4), DOI: 10.9781/ijimai.2018.04.001, 5: 51–59. (accessed January 7, 2021).

Silberztein, M. 2016. *Formalizing Natural Languages: The NooJ Approach*, Cognitive science series, London, UK: Wiley-ISTE.

Silić, J. and I. Pranjković. 2005. *Gramatika hrvatskoga jezika: za gimnazije i visoka učilišta*. Zagreb: Školska knjiga.

Spasić I., J. Livsey, J. A. Keane and G. Nenadić. 2014. Text mining of cancer-related information: Review of current status and future directions. *International Journal of Medical Informatics* 83: 605–623.

Šojat, K., M. Filko and D. Farkaš. 2016. Verbal Multiword Expressions in Croatian. In *Proceedings of the Second International Conference Computational Linguistics in Bulgaria*, Sofia: Institute for Bulgarian Language, Bulgarian Academy of Sciences, 78–85.

Tadić, M. and K. Šojat. 2003. Finding multiword term candidates in Croatian. In *Proceedings of Information Extraction for Slavic Languages 2003 Workshop* (IESL2003), 102–107, Sofija: BAS.

Vidović Bolt, I. 2018. Frazemi - prevoditeljski kamen spoticanja. In *Frazeologija, učenje i poučavanje*. Rijeka: Zbornik radova s Međunarodne znanstvene konferencije.

Wang, Y., L. Wang, M. Rastegar-Mojarad, S. Moon, F. Shen, N. Afzal, S. Liu, Y. Zeng, S. Mehrabi, S. Sohn, et al. 2018. Clinical information extraction applications: A literature review. *Journal of biomedical informatics*. 77: 34–49. DOI: https://doi.org/10.1016/j.jbi.2017.11.011

Wilcox, A. B. and G. Hripcsak. 1998. Knowledge discovery and data mining to assist natural language understanding. In *JAMIA - Journal of the American Medical Informatics Association*, Hanley & Belfus, Inc. Orlando, Annu Fall Symp, 835–839.

Wilcox, A. B. and G. Hripcsak. 2003. The role of domain knowledge in automating medical text report classification. *Journal of the American Medical Informatics Association* 10, no. 4: 330–338.

Zhu, F., P. Patumcharoenpol, C. Zhang, Y. Yang, J. Chan, A. Meechai, et al. 2013. Biomedical text mining and its applications in cancer research. *Journal of Biomedical Informatics* 46: 200–211.

6 HealFavor: Machine Translation Enabled Healthcare Chat Based Application

Sahinur Rahman Laskar,
Abdullah Faiz Ur Rahman Khilji, and
Partha Pakray
Department of Computer Science and Engineering, National
Institute of Technology Silchar, Assam, India

Rabiah Abdul Kadir
Institute of IR4.0, Universiti Kebangsaan Malaysia, Selangor,
Malaysia

Maya Silvi Lydia
Faculty of Computer Science and Information Technology,
Universitas Sumatera Utara, Medan, Indonesia

Sivaji Bandyopadhyay
Department of Computer Science and Engineering, National
Institute of Technology Silchar, Assam, India

CONTENTS

DOI: 10.1201/9781003138013-6

121

6.1   INTRODUCTION

6.1.1   CHATBOT APPLICATION

A chatbot is a well-defined application designed to simulate human-like conversations with users through text/text-to-speech messages. The key task of a chatbot is to help users by providing answers to their questions. Michael Mauldin coined the term "ChatterBot" to express the conversational programs (Mauldin, 1994). A chatbot offers a service powered by rules and sometimes artificial intelligence. This service can be used as an embedded application in popular chat-based systems such as Telegram, WhatsApp, Facebook Messenger, and Imo. There are two types of chatbots; first, function based on rules – it can only respond to very specific commands. Such kinds of chatbots work on rules obtained from a well-defined knowledge base. The availability of such kind of knowledge makes the functioning of the chatbot highly structured and organized. The other type uses machine learning which aims to mimic the brain, and hence hard-coded rules are not required to be fed into the system. Sometimes it can understand language rather than just commands, and it can be designed to continuously get smarter by learning from previous conversations with people and to provide more personalized conversations. Usually, chatbots are created for customer services in various domains like banking, travel agency, or providing assistance in healthcare.

Using chat-based applications provides both operational efficiencies and an additional cost saver to the institute as there is no theoretical limit on the number of parallel chat-based instances with negligible additional costs. This contrasts with the human-based interactions, where every other workforce is associated with a high price to the institution. A significant proportion of customer service is planned to be based on chatbots in the near future (Council Post: Supercharging Your Business With Chatbots Internet, 2021). The following describes the benefits of chatbot:

- Since there is no limit on the number of concurrent requests that such an application can handle, the waiting time is reduced to zero.
- Extended availability, the system is available at all hours, including odd timings throughout the year.
- Chat-based systems provide improved customization and customer engagement.
- The system is easily scalable without any additional costs except server hosting.
- Easier, more central questions are solved with great ease by the system, thereby reducing any costs related to human resources.

The use of chatbots is increasing in various domains or organizations to automate tasks to provide quick assistance or solutions. Applying chatbots to help businesses grow is a recent phenomenon. This chapter focuses on using a chatbot in healthcare, such as improved patient education and treatment compliance.

6.1.2 MOTIVATION

In the medical background, chatbots are utilized for prioritizing patients and guiding people in getting relevant assistance supported by the artificial intelligence method. Chatbots might assist patients who are uncertain of where they must go to get medical care. Most people are unaware of when their conditions need a visit to the doctor and when it is a must to contact a doctor through telemedicine. The general people are now not medically trained to assess the severity of their health issues. This is the place where chatbots can provide an extensive help. A chatbot can acquire relevant facts from patients, and depending on the input, it can supply appropriate data to patients regarding their requirements and recommends further steps.

For the last few years, healthcare chatbots have been adopted more and more. They empower interactive healthcare and offer a window of probability for patients and practitioners to offer ongoing health monitoring and presenting facts rapidly. Health chatbots help individuals with initial screening symptoms and body signs of potential illnesses or diseases. They are designed to guide patients and reveal if they have any severe conditions or diseases. Once the chatbot identifies a severe illness, then the patient is directed to a healthcare professional. Healthcare chatbots can assist in solving general health care issues.

Currently, the ongoing COVID-19 pandemic has affected the healthcare industry significantly. Thus, the sectors are looking at digital platforms to help ease the pressure. AI-enabled chatbots are one technology that can infuse agility in healthcare operations and are fundamental to adopt as the healthcare industry focuses more on patient recoveries. This approach prevents the spread of the virus by curbing the inflow of people into hospitals. Healthcare chatbots can handle an unlimited number of conversations and requests simultaneously with lower operational costs. As hospitals function with infrastructure constraints, deploying chatbots can help to manage patients effectively with limited resources. The following are the advantages of implementing a chatbot in healthcare:

- To help practitioner access the correct information about drug dosage on the go and give the proper guidance to their patients,
- To support the user as nutrition coaching and guide,
- To support the patients who require medical assistance at all times. To remind users to take their medication or contraceptives on time,
- To help to provide patient information related to allergies and check-up reports which help in the time of emergencies,
- To support appointment scheduling by matching patients with relevant clinicians,
- To keep track of patient visits and follow-up visits in case the information is required in the future,

- To support lowering the number of repetitive calls made to the clinic,
- To receive instant diagnosis and medical prescription with the advancement of Artificial Intelligence.

Chatbots are being seen to provide substantial assistance and support to both doctors and patients. Unquestionably, chatbots have a massive potential in the healthcare industry as the technology proves essential to address medical needs efficiently. As artificial intelligence and social connectivity expand, healthcare chatbots will soon become one of the key elements in resolving society's requirements.

Moreover, employing machine translation (MT) in healthcare chatbots can be beneficial to the local community. Since a chatbot system is not available in the Indian languages, we propose a healthcare-based multilingual chatbot for the greater good of the community. In this work, we have considered a low-resource language, Assamese, in the MT environment, and its integration into the healthcare chatbot. The Assamese language is one of the scheduled languages of India and the official language of the Indian state of Assam.[1] To allow healthcare chatbot in the local community of Assam for quick assistance, it is essential to develop such a multilingual healthcare chatbot system.

6.2 THEORETICAL BACKGROUND

6.2.1 MACHINE TRANSLATION

6.2.1.1 Machine Translation Approaches

The need for information that can be easily obtained anywhere and anytime is common in today's computer technology development era. Anyone can get information through the various types of services offered to users via the internet easily. Easy access to documents available on the web written in different languages requires a machine translation. Natural language is a very complex language where words have several meanings. MT is part of natural language processing (NLP) and the sub-field of artificial intelligence, which is a computer application that translates one human language (as a source) to another human language (as its target) automatically. MT is classified into two types of approaches: corpus-based machine translation (CBMT) and rule-based machine translation (RBMT) and as shown in Figure 6.1. The periods of the various MT techniques are presented in Figure 6.2.

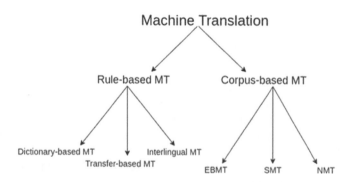

FIGURE 6.1 MT approaches.

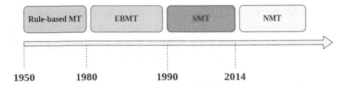

FIGURE 6.2 MT periods.

MT switched from a rule-based (Vauquois 1968; Eisele et al. 2008) to corpus-based approach that leads to developing a language-independent translation system. RBMT relies on a set of rules which needs the intervention of linguistic experts. In corpus-based MT, example-based machine translation (EBMT) (Nagao 1984; Somers 1999) is the naive approach that depends on the text-similarity analogy concept. The EBMT approach translates sentences based on the example sentences translated in advance on the bilingual parallel corpora. However, the EBMT approach has a drawback: we cannot determine various types of sentences by simply relying on other sentences as examples. Another approach included in the corpus-based approach is the statistical machine translation (SMT) (Brown et al. 1990; Koehn 2009). SMT relies on a statistical model that is used to predict target sentences based on source sentences using parameters estimated or learned from the results of parallel corpora analysis. SMT comprises several types, namely word-based translation, phrase-based translation, syntax-based translation, and hierarchical phrase-based translation. Among the four types mentioned above, the most widely used one is phrase-based translation (Koehn et al. 2003). SMT considers the translation task a probabilistic task by predicting the best translation for a given source sentence. The translation and language models are used to evaluate the probability of the target sentence, given the source sentence and the likelihood of the target sentence. And, the decoder is used to find the best translation.

Although phrase-based SMT is better than EBMT and RBMT, it still suffered from a few notable disadvantages, namely, end-to-end problems, long-distance dependency issues, lack of ability to analyse context. It led to the development of Neural Machine Translation (NMT) (Cho et al. 2014; Devlin et al. 2014; Sutskever et al. 2014). At first, NMT appeared together with feedforward neural networks, in which phrase pair scores are calculated by considering a fixed length. Therefore, recurrent neural networks (RNN)-based NMT was introduced (Cho et al. 2014; Sutskever et al. 2014) to tackle variable-length phrases. RNN adopts long short-term memory (LSTM) to capture long-distance dependencies. But the RNN cannot encode all the necessary information when the sentence is too long. The attention mechanism was introduced to handle this issue (Luong et al. 2015; Bahdanau et al. 2016). The disadvantage of RNN is that the input processing follows computation of previous and current words, leading to strict chronological order and lack of parallelization to speed up the learning process. For this reason, transformer model-based NMT was introduced (Vaswani et al. 2017). NMT achieves state-of-the-art performance in both high and low-resource language translation (Laskar et al. 2019a, 2019b, 2019c; Pathak et al. 2019; Pathak and Pakray 2019; Lakew et al. 2020; Lample and Conneau 2019; Wang et al. 2020). In MT, natural languages can be grouped into high-resource, medium-resource, and low-resource languages based on the availability of resources. Most of the languages spoken all over the world are considered low-resource, i.e. they lack sufficiently high number of native speakers, computational data resources, and online data resources.

6.2.1.2   Machine Translation in Healthcare

Medical services are essential, especially for people who live in rural areas. A developing country cannot ignore the health of its people who are living in the countryside. Health services in rural areas are sometimes difficult to obtain, insufficient health workers and inadequate health facilities are difficult to overcome. One approach being developed is to provide long-distance communication for health consulting services via the Internet. It will bridge most medical practitioners or doctors with people in the countryside. However, these practitioners or doctors often do not understand the local languages, so there is a significant communication barrier between the doctor and the patient. Although this situation could benefit from a conventional interpreter, it would undoubtedly take time and money, besides increasing the burden on patients. For overcoming the limitation of human translators and interpreters, Machine Translation (MT) can be beneficial in smaller health organizations with limited resources and few multilingual staff. In MT enabled chatbot, users can use their native language to ask health-related questions, and chatbots can use the same language to help users. This application may be helpful for local people who are not familiar with English. In a multilingual country like India, multilingual healthcare chatbots can help local communities by providing quick help on health-related issues in their language. Figure 6.3 presents one example where a user asks a health-related question in Assamese, and a chatbot replies to the answer in Assamese.

In this work, we have incorporated the low-resource Indian language, Assamese, in the MT module, which acts as an interface for translating Assamese into English for processing text input in the HealFavor application.

6.2.2   SURVEY ON CHATBOTS

6.2.2.1   History and Types of Chatbots

In 1950, Alan Turing posed a question about the definition of thinking and the ability of machines to think (Turing 2009). The Turing test can be performed to reply to this question where a candidate does a "blind" communication with the tested subject (Zemčík 2019). During this test, the thinking ability is identified or replaced with the ability to turn the participant into an intelligent conversation agent. At the end of this test, the candidate can decide whether it is a human being or an artificial agent, i.e. a chatbot.

FIGURE 6.3   MT enabled in HealFavor application.

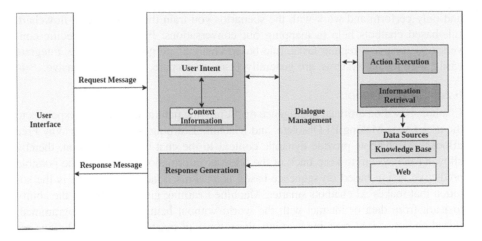

FIGURE 6.4 General chatbot architecture.

A chatbot is a program that serves as an artificial intelligence application in which users or other chatbots communicate with each other. It should be reliable to users in the aspects of social norms, relationships, and accountability. Developers mainly focus on two elements in chatbot development, which are emotions and agency. Artificial intelligence (AI), natural language processing (NLP), and machine learning are technologies used in the chatbot. ELIZA is an early natural language processing computer program built to demonstrate that the communication between man and machine was superficial. ELIZA (Weizenbaum 1966) is designed to imitate a therapist who would ask open-ended questions and even respond with follow-ups. ELIZA is considered to be the first chatbot in the history of computer science. Figure 6.4 shows a general chatbot architecture, where the chatbot can reach out to a broad audience on messaging apps via the user interface.

All chatbots have two essential high-level functions – "understanding" and "answering." These two functions are the most crucial factor when choosing a chatbot for your brand, as they command how much human work is necessary to maintain a chatbot. The design and development of a chatbot involve a variety of techniques. There are two types of chatbot technology, from a series of bots based on rules to AI chatbots. Among these two categories, AI chatbots gained popularity over the recent years.

6.2.2.1.1 Rule-Based Chatbots

A rule-based chatbot is the primary type of chatbot and its versatility in particular. Rule-based chatbots are referred to as decision-tree bots, use a series of defined rules. These rules are the basis for the types of problems the chatbot is familiar with and deliver solutions to. It deals with direct and straightforward question answers. Based on requirements of the user, simple or complicated rules are constructed to develop rule-based chatbots. However, the chatbot would not do anything outside of the defined rules. Therefore, a customer faces difficulty if he/she encounters a new issue that is not included in the program. These chatbots do not learn through interactions and only perform and work with the scenarios you train them for. Like a flowchart, rule-based chatbots help in mapping out conversations. For the case-specific conversation for many regular tasks, rule-based chatbots are good and secure, integrate easily with legacy systems, are generally faster to prepare, and less expensive.

6.2.2.1.2 AI Chatbots

Chatbots based on artificial intelligence or AI provide natural human-like responses to the user (AI is Making BI Obsolete, and Machine Learning is Leading the Way Free eBook). AI chatbots provide dynamic context to the chat-based interaction, thereby allowing the system to keep track of the relevant information and decide the possible predictive action based on sequence-based algorithms. Machine Learning is the solution that makes AI chatbots smarter. Machine Learning gives the chatbot the ability to learn from data or interact with the world without being explicitly programmed. Based on supervised machine learning, three types of machine learning solutions are used, namely input variable (x), an output variable (y), and an algorithm, to learn the mapping function from the input to the output. This technique needs to label conversation data and train a model to learn how to talk with the customer. The other approach of machine learning is based on adaptive unsupervised learning, which consists of input data (x) without corresponding output variables. Unlike supervised learning, the chatbot improves itself by observing data without requiring a teacher or data labelled with correct answers in handling more complex conversations.

AI chatbots that use machine learning predict the context and intent of a question before formulating a response. This type of chatbot generates answers to more complicated questions by garnering the context using advanced NLP techniques. Some other advantages of AI chatbots are learning from information gathered, continuously improving as more data comes in, understanding patterns of behaviour, and having a broader range of decision-making skills. Chatbots provide instant customer support at scale. However, the system poses a challenge in the training phase of its development. The system requires understanding the user's intent when user inputs a piece of text to be queried or an answer to the question posed by the system. Since the user may ask the same question differently, it poses a difficult challenge, particularly for the natural language understanding module of the system. There are three main elements of a chatbot application: dialogue system, natural language understanding, and natural language generation. A dialogue system is responsible for taking the user's input and producing output. A chatbot needs a compatible interface for the interaction between the user and the machine. Natural language understanding is the fundamental unit of a chatbot. Since natural languages are complex, the system must understand the language's lexical, synthetic, and semantic levels. After the chatbot successfully parsed and understood the user's input, it must generate an appropriate response and translate it back to natural language. For this purpose, a natural language generation module is used in a chatbot.

6.2.2.2   Chatbots in Healthcare

Today, healthcare chatbots are not just helping frightened patients assess their symptoms and guide them to receive the appropriate help. The growth of the technology improves the capability of the healthcare chatbots, such as the power of

AI technology in overcoming the challenges of human error and carelessness in healthcare. AI-powered chatbots bring quality and affordable healthcare closer to people. For example, the patients need someone to talk to, listen and guide them on treating their health problems. Instead of providing a better approach to the patients, healthcare chatbots with the emergence of AI have transformed the conversation between patients and doctors and the management of hospitals and healthcare centres.

The healthcare industry started with a telehealth system that comprises the dissemination of healthcare services and information extraction from electronic devices and telecommunication technologies. Telehealth is helpful for communication, such as answering queries and providing recommendations to patients and doctors in certain circumstances. It is available 24/7; however, a full-fledged schedule becomes difficult for the doctors to answer each call or message from the patients or users. In this regard, chatbots can quickly answer relevant questions from the patients or users regarding the information about drug dosage, nutrition coaching, medical assistance, etc., at any time. Chatbots also gather users' data to make an appointment with the doctor.

Implementation of a combination of AI and chatbots are worthful in a healthcare organization. There are three main considerations. First, provide exceptional patient experiences and understand and engage the users in online conversation by offering guidance and personalized recommendations. Second, in particular time duration, the chatbots become smarter with a great understanding of contextual information by applying machine learning and other relevant technologies. Finally, a combination of AI and chatbots will be intelligent in determining entitled problems, which will only succeed when accessing and gathering accurate data and information (How Chatbot Technology Revolutionize the Healthcare Industry? | by ChatbotNews | Chatbot News Daily). Theoretically, the combination of AI and chatbots contributes to the healthcare industry and is expected to transform the sector to become more efficient with the help of automated tools to gather collective information from reliable sources.

6.3 DATASET

To prepare suitable data for the prototype system of HealFavor (Khilji et al. 2020b) we have collected from various online sources namely Drugs.com[2], WebMD[3], Stanford Medicine[4] and undertaken the necessary steps to ensure the authenticity of data in our previous works (Khilji et al. 2020a, 2020b). Additionally, common symptoms of highly infectious diseases such as coronavirus disease (COVID-19) (Mehta et al. 2020) and necessary precautionary measures are included. In this work, we are concerned about the Assamese parallel data of our English version dataset since the goal is the English–Assamese MT-enabled chatbot system. We have performed MT work on English–Assamese pair in (Laskar et al. 2020), and the baseline results are shown in Table 6.1. In Table 6.1, we have reported the bilingual evaluation understudy (BLEU) score (Papineni et al. 2002; Laskar et al. 2020) along with other automatic evaluation metrics, namely, metric for evaluation of translation with explicit ordering (METEOR) (Lavie and Denkowski 2009), F-measure,

TABLE 6.1
Baseline system results of English–Assamese MT

Translation	System	BLEU	METEOR	F-Measure	RIBES	TER (%)	WER (%)
English to	SMT (PBSMT)	3.43	0.138945	0.405864	0.185537	93.12	93.31
Assamese	NMT (RNN)	5.55	0.183634	0.482956	0.252370	91.64	92.27
Assamese to	SMT (PBSMT)	4.54	0.045206	0.090317	0.121474	106.14	107.00
English	NMT (RNN)	7.72	0.094548	0.228664	0.300727	89.93	91.53

rank-based intuitive bilingual evaluation score (RIBES) (Isozaki et al. 2010), translation edit rate (TER) (Snover et al. 2006) and word error rate (WER) (Morris et al. 2004). English–Assamese MT (Laskar et al. 2020) is in the beginning stage. BLEU score is low, but we mentioned here that we have performed manual translation for medical domain data. Therefore, the current translation models are not apt for translating texts in the medical domain, primarily for the concerned language pair, i.e., English to Assamese and Assamese to English. For this reason, we have performed manual translation of medical domain data from English to Assamese to avail the medical domain parallel data required to obtain better translation models. The sample dataset of such multi-lingual data will be available at: https://github.com/cnlp-nits/HealFavor.

6.4 SYSTEM DESCRIPTION

6.4.1 NEURAL MACHINE TRANSLATION MODEL

We need to enable MT models of low-resource English to Assamese and vice-versa in our chatbot system in this work. For this reason, we have implemented NMT-based transformer models by utilizing our dataset, as mentioned in Section 6.3. We have chosen transformer model-based NMT since it attains state-of-the-art performance in the MT domain. The primary transformer model comprises six layers in the encoder and six layers in the decoder. Each encoder and decoder layer have an identical architecture, and consist of two sub-layers: self-attention and a feed-forward neural network. By self-attention, the encoder can view other words in the input sentence when encoding a specific word. It helps to solve parallelization. The decoder has an attention layer between both layers that helps focus on relevant segments of the input sentence. There are six basic steps to calculate self-attention. First, calculate three vectors: query, key, and value from each of the encoder's input vectors (embeddings of each word). It is calculated by multiplying the embedding during the training process. Secondly, calculate a self-attention score for each word of the input sequence. It determines how much importance to give to other segments of an input sentence while encoding a particular word. It is calculated using the dot product of the query vector with the key vector of the considering word. In the third and fourth steps, the computed self-attention score is divided by the square root of the dimension of the key vectors. It helps to obtain more stable gradients. Then the score is normalized via a softmax operation. This obtained score determines the importance of each word at the respective position. The fifth step is used to keep intact values of the focus words, which acquire via multiplying the softmax score with each value vector. Then sum up the weighted value vectors in the last step. Moreover, the transformer model uses multi-head attention rather than one dimension. The transformer model contains a feed-forward neural network used to add positional encoding while encoding each word since word position is relevant to the translation.

The training, testing, validation of the translation models is undertaken using the OpenNMT-py[5] toolkit of PyTorch, which is freely available (Laskar et al. 2020). The experiment is conducted separately in two directions (English to Assamese and Assamese to English translation) through four main steps: preprocessing, unsupervised pretraining, supervised training, and testing. In the preprocessing step, we have tokenized source-target parallel sentences and obtained a vocabulary dimension of 50,000. It is required to perform indexing of words during the training process. We have utilized Glove[6] (Pennington et al. 2014) unsupervised word embeddings on monolingual data to capture semantic word relationships. The pretraining is carried out up to 100 iterations with embedding vector size 200. The transformer model is adopted for the training step following the default configuration of six layers, eight attention heads, and 0.1 drop-outs. Also, we have used Adam optimizer with the default learning rate of 0.001. The models are trained on a single GPU NVIDIA Quadro P2000 GPU up to 50,000 epochs, and obtained optimum trained model is tested on the test data (Laskar et al. 2020) to check the model's performance. To find out the best translations, we have used the beam search technique with default size 5. We have evaluated the test data (Laskar et al. 2020) using the BLEU metric and obtained BLEU scores of 5.93, 8.91 for English to Assamese and Assamese to English translation. The difference between the present and previous work is that the transformer model is built in this work, unlike the RNN model (Laskar et al. 2020).

6.4.2 HEALFAVOR

In this section, we discuss the system architecture and the prototype system.

6.4.2.1 System Architecture

To allow for easy access to the healthcare facilities for the local population and scale our system, we have developed our chat-based system. The backend framework of our system follows that of Rasa (Bocklisch et al. 2017). The underlying model also leverages the embedding dialog policy based on recurrent networks (Bocklisch et al. 2017). The algorithm used for this purpose is the FAIR algorithm or the Facebook artificial intelligence research-based algorithm (Wu et al. 2017). Here, embeddings are used by neural networks to solve a variety of downstream tasks. Cosine similarity is used for similarity calculations with the probable states.

Previous works on similar datasets (Khilji et al. 2020a, 2020b) have employed bag of words (BoW) representation. It was used in order to leverage the features from those vectors. These vectors were then passed through the embedding layer. When the word tokens are converted into an n-dimensional matrix as a resultant of the embedding layer, this is then fed into the network's dense layer, allowing the model to predict and map it to the most probable output from preset output definitions. To calculate the relevant output from the model, the recurrent network (Mikolov et al. 2011) collects messages from patients and past input messages and calculates attention from these sentences (Vaswani et al. 2017). The recurrent layer of the model obtains collective inputs from both the embedding and attention layer neurons. The output of this layer is then passed to a separate embedding layer. The dialogue generation module then uses the vectors obtained from the previous system-generated response and the output of the final layer to generate patient responses. The LSTM states of the model are multiplied with attention probabilities randomly element-wise to accommodate the random steps. The diagram in Figure 6.5 summarizes the proposed system architecture. To calculate the loss required to adjust the model weights in the backpropagation step, we have used the following loss functions.

6.4.2.2   The Prototype Chatbot System

We curated to the full extent we have built a robust multilingual chat-based system to interact with the users for leveraging the dataset. It will also allow the authors of the paper to evaluate the model in the real working environment and know the system's robustness. A working example snapshot of our system is shown in Figure 6.6, and the example shows how the patient interacts with the system to book an appointment with a physician. The back end of the system is used together with the Mongo database (MongoDB) to better serve, store, and analyse user data to provide customized services for patients (data will only be collected and stored when the user provides permission and the stored information is stored and will not be shared without proper consent). This chat data can also be used to analyse the chat centrally and allow the concerned authorities to handle outbreaks at an efficient and fast pace. To allow the conversation to proceed in the local language and analyse and process and provide the output in the same language, we have used an NMT model in the intermediary step of our model. It provides us with a dual advantage. One visible advantage is the availability of the conversation in the local language, Assamese in our case. Another advantage is that all the processing and analysis of the data is done in the English language, thus enabling us to centrally analyse, collect and store the dataset contributing to the improvement of the system. For typing the Assamese input texts, users can use any freely available Assamese keyboards. For our use case, we have used the keyboard layout as provided by Xobdo[7] and is available for both Windows[8] and macOS[9]. The NMT model used for translation is as discussed in Section 6.4.1.

6.5   DISCUSSION

To automatically evaluate such a system's efficiency, there is a need to develop an automatic evaluation technique. The main difference between the present work with the previous work is that it supports an additional local language i.e., Assamese together with English. Also, this work includes additional data including, but not limited to the COVID-19 pandemic. The contribution of multi-lingual features in the healthcare chatbot helps the local community and enhances the use of

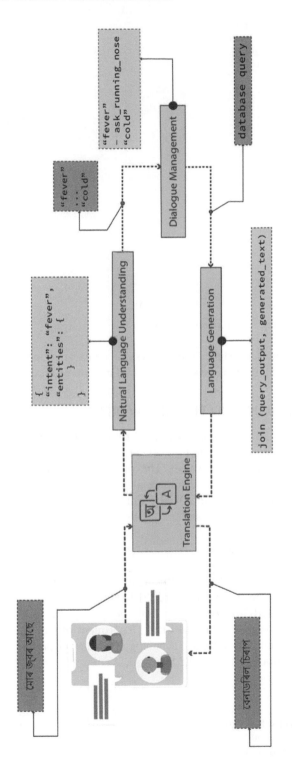

FIGURE 6.5 System architecture diagram.

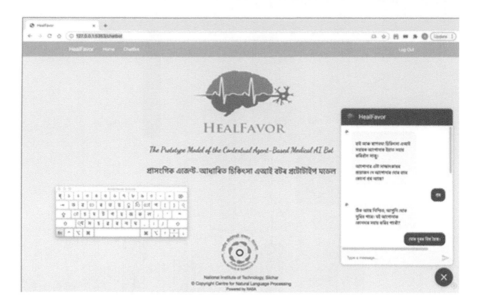

FIGURE 6.6   Working system prototype demo.

technology to a greater extent. Such a system gives an impetus to the use of technology by the indigenous people of the Indian state of Assam.

6.6   CONCLUSION AND FUTURE WORKS

Our work proposed a multilingual chat-based user agent for the healthcare domain and developed a prototype system to leverage the English–Assamese NMT model and the chat-based healthcare system. The presented system can be seen as an extension of our previous work (Khilji et al. 2020a, 2020b), with the Assamese language support being the main novelty. Future work will include adding more low-resource languages into the system, facilitating an increase in the dataset.

ACKNOWLEDGEMENT

This work is supported by the Department of Science & Technology (DST) and Science and Engineering Research Board (SERB), Govt. of India and the Research Project Grant No. CRD/2018/000041. Authors would like to thank Center for Natural Language Processing (CNLP) and Department of Computer Science and Engineering at National Institute of Technology, Silchar for providing the requisite support and infrastructure to execute this work.

NOTES

1 http://www.minorityaffairs.gov.in/sites/default/files/2.%2052nd%20Report%20English.pdf
2 https://www.drugs.com/
3 https://www.webmd.com/

4 http://med.stanford.edu/
5 https://github.com/OpenNMT/OpenNMT-py
6 https://github.com/stanfordnlp/GloVe
7 http://www.xobdo.org/
8 http://www.xobdo.org/article/win7
9 http://www.xobdo.org/article/mac

REFERENCES

AI Is Making BI Obsolete, and Machine Learning Is Leading the Way Free eBook. https://www.explorium.ai/resource/ai-is-making-bi-obsolete-and-machine-learning-is-leading-the-way/download/

Bahdanau D., K. Cho, and Y. Bengio. 2016. Neural machine translation by jointly learning to align and translate. arXiv:14090473 [cs, stat].

Bocklisch T., J. Faulkner, N. Pawlowski, et al. 2017. Rasa: Open source language understanding and dialogue management. arXiv:171205181 [cs].

Brown P. F., J. Cocke, S. A. Della Pietra, et al. 1990. A statistical approach to machine translation. *Computational Linguistics*. 16: 79–85.

Cho K., B. Van Merriënboer, C. Gulcehre, et al. 2014. Learning phrase representations using RNN encoder-decoder for statistical machine translation. arXiv preprint arXiv:14061078

Council Post: Supercharging Your Business With Chatbots [Internet]. 2021. [cited January 25]. Available from: https://www.forbes.com/sites/forbesagencycouncil/2019/04/10/supercharging-your-business-with-chatbots/?sh=ab090fa7b8ec

Devlin J., R. Zbib, Z. Huang, et al. 2014. Fast and robust neural network joint models for statistical machine translation. Proceedings of the 52nd annual meeting of the Association for Computational Linguistics (Volume 1: Long Papers). pp. 1370–1380.

Eisele A., C. Federmann, H. Saint-Amand, et al. 2008. Using Moses to integrate multiple rule-based machine translation engines into a hybrid system. Proceedings of the Third Workshop on Statistical Machine Translation. Association for Computational Linguistics. pp. 179–182. https://aclanthology.org/W08-0328/

How Chatbot Technology Revolutionize the Healthcare Industry? | by ChatbotNews | Chatbot News Daily. https://chatbotnewsdaily.com/how-chatbot-technology-revolutionize-the-healthcare-industry-957d8b700001

Isozaki H., T. Hirao, K. Duh, et al. 2010. Automatic evaluation of translation quality for distant language pairs. Proceedings of the 2010 Conference on Empirical Methods in Natural Language Processing. Association for Computational Linguistics. pp. 944–952. https://aclanthology.org/D10-1092

Khilji A. F. U. R., S. R. Laskar, P. Pakray, et al. 2020a. HealFavor: A chatbot application in healthcare. *Analysis of Medical Modalities for Improved Diagnosis in Modern Healthcare.*CRC Press. https://www.taylorfrancis.com/chapters/edit/10.1201/9781003146810-3/healfavor-chatbot-application-healthcare-abdullah-faiz-ur-rahman-khilji-sahinur-rahman-laskar-partha-pakray-rabiah-abdul-kadir-maya-silvi-lydia-sivaji-bandyopadhyay

Khilji A. F. U. R., S. R. Laskar, P. Pakray, et al. 2020b. HealFavor: Dataset and a prototype system for healthcare chatbot. 2020 International Conference on Data Science, Artificial Intelligence, and Business Analytics (DATABIA). IEEE. pp. 1–4. https://ieeexplore.ieee.org/abstract/document/9190281

Koehn P. 2009. *Statistical Machine Translation*. Cambridge University Press.

Koehn P., F. J. Och, and D. Marcu. 2003. *Statistical Phrase-Based Translation*. University of Southern California Marina Del Rey Information Sciences Inst.

Lakew S. M., M. Negri, and M. Turchi. 2020. Low resource neural machine translation: A benchmark for five African languages. arXiv:200314402 [cs].

Lample G., and A. Conneau. 2019. Cross-lingual language model pretraining. arXiv: 190107291 [cs].

Laskar S. R., Dutta A., Pakray P., et al. 2019a. Neural machine translation: English to Hindi. 2019 IEEE Conference on Information and Communication Technology. IEEE. pp. 1–6.

Laskar S. R., P. Pakray, and S. Bandyopadhyay. 2019b. Neural machine translation: Hindi-Nepali. Proceedings of the Fourth Conference on Machine Translation (Volume 3: Shared Task Papers, Day 2). pp. 202–207.

Laskar S. R., R. P. Singh, P. Pakray, et al. 2019c. English to Hindi multi-modal neural machine translation and hindi image captioning. Proceedings of the 6th Workshop on Asian Translation. Association for Computational Linguistics. pp. 62–67. https:// aclanthology.org/D19-5205

Laskar S. R., A.F.U.R. Khilji, P. Pakray, et al. 2020. EnAsCorp1.0: English-Assamese Corpus. Proceedings of the 3rd Workshop on Technologies for MT of Low Resource Languages. Suzhou, China: Association for Computational Linguistics. pp. 62–68.

Lavie A., and M. J. Denkowski. 2009. The METEOR metric for automatic evaluation of machine translation. *Machine Translation*. 23: 105–115.

Law E. L.-C., V. Roto, M. Hassenzahl, et al. 2009. Understanding, scoping and defining user experience: A survey approach. Proceedings of the SIGCHI Conference on Human Factors in Computing Systems. ACM. pp. 719–728. https://doi.org/10.1145/1518701.1518813

Luong M.-T., H. Pham, and C. D. Manning. 2015. Effective approaches to attention-based neural machine translation. arXiv:150804025 [cs].

Mauldin M. L. 1994. ChatterBots, TinyMuds, and the Turing test: Entering the Loebner prize competition. Proceedings of the 12th National Conference on Artificial Intelligence. Seattle, WA, USA, July 31 - August 4, Volume 1, pp. 16–21.

Mehta P., D. F. McAuley, M. Brown, et al. 2020. COVID-19: Consider cytokine storm syndromes and immunosuppression. *Lancet (London, England)*. 395: 1033.

Mikolov T., S. Kombrink, L. Burget, et al. 2011. Extensions of recurrent neural network language model. 2011 IEEE International Conference on Acoustics, Speech and Signal Processing (ICASSP). IEEE. pp. 5528–5531.

Morris A. C., V. Maier, and P. Green. 2004. From WER and RIL to MER and WIL: Improved evaluation measures for connected speech recognition. Eighth International Conference on Spoken Language Processing, ISCA, http://www.isca-speech.org/ archive/interspeech_2004/i04_2765.html

Nagao M. 1984. A framework of a mechanical translation between Japanese and English by analogy principle. *Artificial and Human Intelligence*. 351–354. https://dl.acm.org/doi/ 10.5555/2927.2938

Papineni K., S. Roukos, T. Ward, et al. 2002. Bleu: A method for automatic evaluation of machine translation. Proceedings of the 40th annual meeting of the Association for Computational Linguistics. Association for ComputationalLinguistics, pp. 311–318. https://aclanthology.org/P02-1040

Pathak A., and P. Pakray. 2019. Neural machine translation for indian languages. *Journal of Intelligent Systems*. 28: 465–477.

Pathak A., P. Pakray, and J. Bentham. 2019. English–Mizo machine translation using neural and statistical approaches. *Neural Computing and Applications*. 31: 7615–7631.

Pennington J., R. Socher, and C. Manning. 2014. GloVe: Global vectors for word representation. Proceedings of the 2014 Conference on Empirical Methods in Natural Language Processing (EMNLP). Doha, Qatar: Association for Computational Linguistics. pp. 1532–1543.

Shawar B. A., and E. Atwell. 2007. Different measurement metrics to evaluate a chatbot system. Proceedings of the Workshop on Bridging the Gap: Academic and Industrial Research in Dialog Technologies, Association for ComputationalLinguistics, pp. 89–96. https://aclanthology.org/W07-0313

Snover M., B. Dorr, R. Schwartz, et al. 2006. A study of translation edit rate with targeted human annotation. Proceedings of the 7th Conference of the Association for Machine Translation in the Americas: Technical Papers, Association for Machine Translation in the Americas, pp. 223–231. https://aclanthology.org/2006.amta-papers.25

Somers H. 1999. Example-based machine translation. *Machine Translation*. 14: 113–157.

Sutskever I., O. Vinyals, and Q. V. Le. 2014. Sequence to sequence learning with neural networks. *Advances in Neural Information Processing Systems*. pp. 3104–3112. https://proceedings.neurips.cc/paper/2014/hash/a14ac55a4f27472c5d894ec1c3c743d2-Abstract.html

Turing A. M. 2009. Computing machinery and intelligence. In *Parsing the Turing Test: Philosophical and Methodological Issues in the Quest for the Thinking Computer*, editors. R. Epstein, G. Roberts, and G. Beber, 23–65, Dordrecht: Springer Netherlands.

Vaswani A., N. Shazeer, N. Parmar, et al. 2017. Attention is all you need. *Advances in Neural Information Processing Systems*, Curran Associates, Inc. pp. 5998–6008. https://proceedings.neurips.cc/paper/2017/file/3f5ee243547dee91fbd053c1c4a845aa-Paper.pdf

Vauquois B. 1968. A survey of formal grammars and algorithms for recognition and transformation in mechanical translation. *IFIP Congress*, Volume 2, (2). pp. 1114–1122. Information Processing, Proceedings of {IFIP} Congress 1968, Edinburgh, UK, 5-10 August 1968, Hardware, Applications.

Wang X., Y. Tsvetkov, and G. Neubig. 2020. Balancing training for multilingual neural machine translation. arXiv:200406748 [cs].

Weizenbaum J. 1966. ELIZA—A computer program for the study of natural language communication between man and machine. *Commun ACM*. 9: 36–45.

Wu L., A. Fisch, S. Chopra, et al. 2017. StarSpace: Embed all the things! arXiv:170903856 [cs].

Zemčík M. T. 2019. A brief history of chatbots. *DEStech Transactions on Computer Science and Engineering*. http://dpi-journals.com/index.php/dtcse/article/view/31439

7 Development of a Machine Translation System for Promoting the Use of a Low Resource Language in the Clinical Domain: The Case of Basque

Xabier Soto, Olatz Perez-de-Viñaspre,
Maite Oronoz, and Gorka Labaka
HiTZ Basque Center for Language Technologies – Ixa,
University of the Basque Country UPV/EHU, Donostia, Spain

CONTENTS

DOI: 10.1201/9781003138013-7

7.1   INTRODUCTION

Basque is a pre-Indo-European language spoken in the Basque Country, a region spanning an area in northeastern Spain and southwestern France, with a population of around 3 M people.[1] Nowadays, the Basque language is spoken by 28,4% of the people in all territories (751,500 active speakers and 1,185,500 passive). Of these active speakers, 700,300 live in the Spanish part (Basque Country and Navarre autonomous communities) and the remaining 51,200 live in the French part ('Euskal Hirigune Elkargoa' in Basque, or 'Communauté d'agglomération du Pays Basque' in French).[2]

We can then state that Basque is a minority language that persists surrounded by two powerful languages, Spanish and French. Linguistically, Basque is considered an isolated language of unknown origin, and so, it does not share any characteristics with its neighboring Romance languages. In the following examples the reader may recognize the distance among these languages, along with the English glosses for reference:

Basque

buru-ko	*mina*	*dauka*	*eta*	*hiru*	*egun*	*daramatza*	*botaka.*
head-GEN	pain	has	and	three	days	carries	vomiting.

'he/she has a headache and has been vomiting for three days.'

Spanish

le		*duele*	*la*	*cabeza*	*y*	*lleva*	*tres*	*días*	*vomitando.*
to-him/her		hurts	the.F	head	and	carries	three	days	vomiting.

'he/she has a headache and has been vomiting for three days.'

French

il/elle	*a*	*un*	*mal-de-tête*	*et*	*vomit*	*depuis*	*trois*	*jours.*
he/she	has	a	headache	and	vomits	since	three	days.

'he/she has a headache and has been vomiting for three days.'

Nowadays, Basque holds co-official language status with Spanish in the Basque Autonomous Community and in the northern part of Navarre, but for centuries Basque has not been an official language; it was out of educational systems, out of media, and out of industrial environments, being its use mostly limited to the private sphere. It was not until the late 1960s that the standardization process of the Basque language started, while the current autonomous institutions, including the health-related ones, were created in the early 1980s. Due to these and other features, the use of Basque language in the bio-sanitary system is still low.

In the Basque Autonomous Community, the oral communication between patients and healthcare workers (especially nurses and physicians) can be done in Basque given that both sides are Basque speakers, but almost all of the reports are still written in Spanish. This is often caused by a lack of habit of the healthcare workers to write in Basque, owing to diverse reasons like the language used in education, knowledge of health-related terminology, etc. However, even when the healthcare workers want to write the reports in Basque, the diglossic situation, where everyone knows Spanish and only a minority knows Basque, makes them feel forced to write the documents in Spanish. One of the reasons for this is that Osakidetza (the public health service in the Basque Autonomous Community) has a centralized system for storing the health information of the patients, so healthcare workers can access the clinical records of the patients, which have been possibly written by another colleague. Thus, when a healthcare worker wants to write a report in Basque, if the following readers are not Basque speakers, the safety of the patient can be put at risk.

For comparison[3], we have studied how the communication between patients and healthcare workers is done in other multilingual countries. In Canada, in the areas where more than one language is official, patients decide the language they want to use in their communications (Desjardins 2003), so all their health records are written in that language. By contrast, in Belgium, the communities are separated by the language they use, each of them having their own public health service. In the case of Brussels, being a bilingual area, health services are offered both in French and Flemish (Gerkens and Merkur 2010). As a last example, in Luxembourg, where German, French, Italian, English, and Portuguese are all widely used languages, they use French as lingua franca for communications in the health care context (European Observatory on Health Care Systems 1999). In the Basque Country, the language communities are merged, using Spanish as lingua franca in the southern part of the Basque Country and French in the northern part. In this situation, the linguistic rights of the Basque-speaking patients and healthcare workers are not preserved, since, as said before, even if the oral communication can be done in Basque, the whole clinical attention can not be given in Basque. This work summarises the first steps done for creating the necessary conditions for Basque speaking healthcare workers to write their reports in Basque, while guaranteeing that the safety of the patients is not put at risk.

7.2   ITZULBIDE

Itzulbide is a project defined in cooperation between Osakidetza and the University of the Basque Country that aims to promote the use of Basque in EHRs. As mentioned before, since the EHR storage system in Osakidetza is centralized, any next doctor to see the patient may access all the previous EHRs. In the case of monolingual speakers of Spanish, they may not understand what is written in Basque. This project wants to take steps to address this problem, creating an MT system from Basque to Spanish for the clinical domain.

In order to train the automatic translator, it is essential to collect parallel medical reports in Basque and Spanish. To this end, work has been done on a web application that helps healthcare workers to write bilingual EHRs.

In sub-section 7.2.1 we make a general description of the Itzulbide project, in sub-section 7.2.2 we specify how the compiled documents are classified, and in sub-section 7.2.3 we briefly present the web application designed for collecting the bilingual corpus to be used by the MT system.

7.2.1 PROJECT DESCRIPTION

One of the main challenges of Osakidetza is to progressively increase the presence of Basque in clinical documentation and medical records. To this end, Osakidetza proposed a public bid which the HiTZ group won and later became the Itzulbide project.

There is a growing demand from Osakidetza's healthcare workers for the use of the Basque language in medical records, but this does not materialize as they are aware that a large part of the professionals does not speak both languages. In addition, the progressive increase in the use of Basque in oral communications between professionals and between professionals and patients leads to a natural increase in the use of Basque in the written section. However, in practice, it happens that it requires an additional effort on the part of the professional, who in the context of a consultation that is being carried out in Basque, has to transcribe its content in Spanish in order to guarantee the continuity of the patient's care. This, in turn, does not encourage the professional to use the patient's preferred language.

The moment to face this project in 2018 was considered adequate for two main reasons:

1. 46.60% of active employees in structural positions and 56.29% of the temporary staff were bilingual at that moment.
2. The paradigm shift that has involved the use of neural networks and deep learning in Natural Language Processing (NLP) in general, and in machine translation in particular, changed, obtaining very high-quality translations.

Although there has been an accelerated growth in the development of machine translation systems, these do not cover the specific needs that may arise in the field of healthcare. There are no tools that respond to the Osakidetza healthcare context with the required reliability, accuracy, and quality.

Taking into account all these facts, the development of a specific MT tool was considered essential to maximize the accuracy and reliability of text translated from Basque into Spanish based on a bilingual corpus of clinical texts.

Through this project, we contribute to Osakidetza, not only by responding to the strategic challenges set out in its linguistic policy, but also facing a growing social demand. The quality of care is increased by improving the communication between professionals and patient-professionals and enhancing their satisfaction. The cost/benefit ratio is also improved as external translation costs are reduced.

The project is divided into three phases:

1. Corpus creation: volunteer health care workers are recruited to write bilingual EHRs (Basque/Spanish).

2. MT training: using the bilingual corpus and other available bilingual and monolingual resources, the neural machine translator is trained.

3. MT evaluation: part of the volunteers manually evaluate the system.

7.2.2 CLASSIFICATION OF THE MEDICAL RECORDS

Classifying the documents considering different aspects will help us to perform diverse experiments when training and evaluating the neural translator. The web application for compiling the bilingual corpus considers the characteristics described in the following lines.

Volunteers participating in the project will need to create an account in the web application designed for the corpus collection. When logging in for the first time, each healthcare worker will indicate the health organization in which they work. Osakidetza is organized in 'integrated health organizations', known by their acronym in Spanish OSI (*'Organización de Servicio Integrado'*), that aim to integrate the different levels of patient care, and group hospitals and health centers in the same organization. They are organized by geographical areas. For example, the OSI in the region around Donostia is composed of 32 health centers and outpatient clinics, and a university hospital. The scope of action is a reference population of nearly 400,000 inhabitants near the capital of the province of Gipuzkoa, Donostia. All the health organizations in Osakidetza are listed in the web application[4].

When creating their account, the volunteer healthcare worker can also indicate whether they are a nurse, a doctor, or have any other position, and will mark their specialty or area of care. Even if the healthcare worker indicates his/her specialty, the documents are not displayed considering this specialty, but the one that has been assigned to each pair of documents. We must consider that, for example, some doctors can make extra turns in different specialties.

Once the volunteers' details have been indicated, for each pair of reports, the specialty and the document type are indicated. The considered specialties and document types are presented in sub-section 7.3.1, and listed in the first column of Table 7.3 and Table 7.2 respectively.

7.2.3 WEB APPLICATION FOR CORPUS COLLECTION

For collecting a proper corpus, it is important to provide adequate guidelines to the users. With this aim, the volunteers working on the elaboration of the bilingual corpus were asked to follow these guidelines:

- Align the source and target text at the sentence level. To do so, write each sentence in one line. If a source sentence needs more than one target sentence to translate, include them in the same line, and vice versa.
- Use the most natural way of writing as possible. We need a corpus the most realistic as possible, and so it should be the language used[5].
- Use a formal style; lexical variations related to dialectal use are accepted, but not orthographical variations that deviate from standardized terms.

Apart from providing adequate guidelines, one of the aims of the web application is to reduce the effort the volunteers may do writing the records. For that purpose, we included in the web application some tools that may be of help when collecting the bilingual corpus:

- Access to a set of 48 reference discharge records of different specialties in Basque (Joanes Etxeberri Saria V. Edizioa 2014).
- A dynamic bilingual dictionary which the participants can enrich with their collaborations, so anyone can check the proposals of their mates.
- An integrated medical dictionary in the Basque writing text area, that whenever the volunteer types the colon symbol, it searches the string typed next in a specialized dictionary. Thus, users avoid checking in a different tab for the source word in a digital dictionary, and speeds up the writing process.

In addition, the biggest deal in collecting a parallel corpus is the alignment of the source and target sentences. Even if the guidelines in this respect were clear, it should be considered that sometimes a sentence in one language needs more than one sentence in the other language. This may be faced using an automatic sentence splitter, but then some errors could inevitably be introduced. Trying to minimize these, we developed some tools to perform the sentence mapping, by a) splitting the source text when copying to the app, having a sentence by line; and b) underlining with colors the corresponding source and target sentence as seen in Figure 7.1.

Snapshot of the web application for corpus collection. The screen is divided into two vertical panes, with a text box for writing the health record in Spanish in the bottom left (under the text Gaztelania, meaning Spanish language in Basque) and an equal-sized text box in the bottom right for writing the equivalent document in Basque (under the text Euskara, meaning Basque language in Basque). Aligned sentences corresponding to the current cursor position are highlighted in different colors for each language. Centered at the top left quarter there is a menu for choosing the document specialty (preceded by the text Espezialitatea, meaning specialty in Basque), while a menu for choosing the document type appears centered at the top right (after the text Txosten mota, meaning document type in Basque). In the top left of the screen there is a menu with words in Basque for accessing other sections of the web application.

To conclude this section, we would like to highlight two issues concerning the corpus collection: i) the way in which the collection process has been organized and ii) the fact that patient privacy is always guaranteed.

Each OSI has had a project manager who has listed the names of the bilingual healthcare workers who might be interested in participating in Itzulbide. Both the institutional representatives of the health system and the technical managers, and some of the authors of this chapter, have gone (before the COVID-19 pandemic) to some of these centers to make a presentation of the project, explain the technical part of the corpus collection application and solve professionals' doubts. During the pandemic period, this presentation has been recorded in the form of a video and made available to interested sanitary workers. Each OSI decides whether or not to

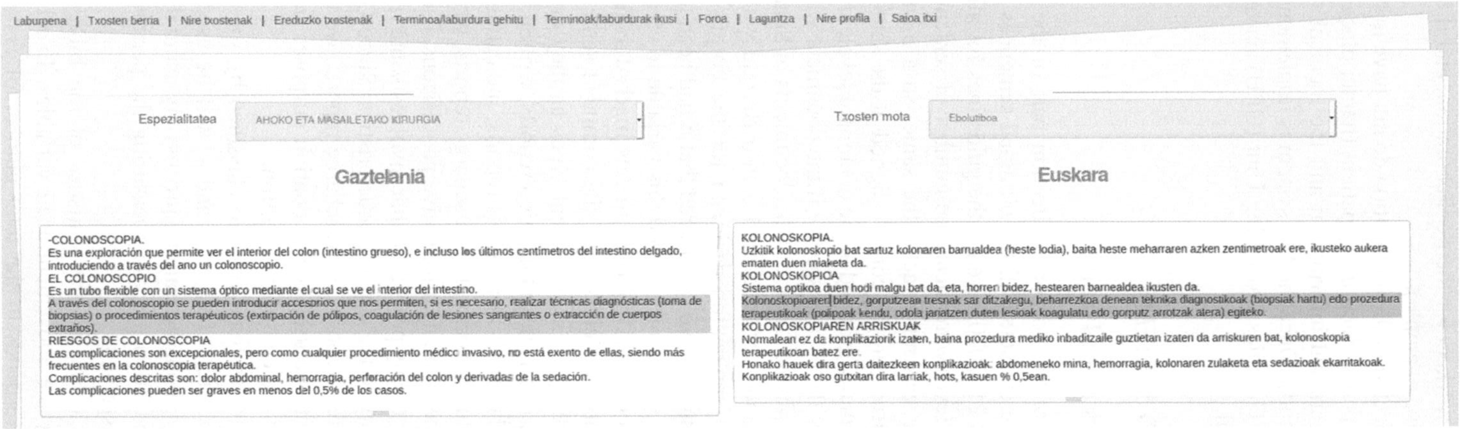

FIGURE 7.1 Snapshot of the web application for corpus collection, highlighting in colors the aligned sentences in Spanish (left, under '*Gaztelania*') and Basque (right, under '*Euskara*'), and showing the menus for specialty ('*Espezialitatea*') and document type ('*Txosten mota*').

compensate the healthcare workers for their collaboration, and if this is the case, in what way (with days off, for example).

The medical reports collected in the application do not have to be real, but if they are, they do not contain personal data of the patient (name, place of birth, etc.). In Osakidetza, the patient's personal information and the medical reports are stored separately and linked by a code. Once the MT system is implemented in production, it will run on Osakidetza's Graphics Processing Unit (GPU) servers, so the flow of information will be internal.

7.3 RESOURCES AND SYSTEMS

7.3.1 BILINGUAL CORPUS FROM ITZULBIDE

As an intermediate result of the corpora collection part of the Itzulbide project described in section 7.2, we extracted all the bilingual sentences introduced in the web application until April 21, 2020. Even if the guidelines for aligning the source and target text at sentence level were clear, the number of sentences automatically extracted in each language was not exactly the same, so a manual revision was performed until we had a true parallel corpus aligned at the sentence level.

The sentences were originally grouped at the document level, and each document had a variable controlled by the user to specify if the document was finished or not. In the most usual scenario, physicians wrote the documents in Spanish as they were working with the patient, and later translated it into Basque with the help of the tools available in the web application designed for the corpus collection[6]. If a document was marked as finished, we considered that it had been properly reviewed by the doctor who wrote it, so we further used this variable to create the evaluation corpus only with sentences coming from documents marked as finished.

Another important variable for classifying the compiled sentences is the document type. We defined five different document types as they are commonly distinguished in clinical scenarios: 1) hospitalization reports, written when the patients are initially derived to the hospital; 2) progress reports, indicating the evolution of the patients while they are in the hospital; 3) discharge reports, written at the end of the stay of a patient in the hospital; 4) informative permissions, used when a patient has to go through a surgery or any procedure that involves some risk; and 5) others, for documents not filling any of the above characteristics. Since Osakidetza's main priority is to translate progress reports and discharge reports, we used this variable for selecting the sentences to be used for evaluation purposes in one of our evaluation scenarios defined at the end of this section.

Finally, the documents in Itzulbide are also classified by specialty (e.g.: emergencies, nursing, pediatrics, etc.). This information will be used in some of our defined scenarios for helping both the training process, by using distinctive tags for each specialty, and the evaluation process, measuring the performance of the designed systems in a given specialty.

Tables 7.1, 7.2, and 7.3 sum up the statistics of the bilingual corpus from Itzulbide, disaggregated by the abovementioned variables. In all these tables, the number of documents corresponds to the total number of compiled documents,

TABLE 7.1

Statistics of Itzulbide bilingual corpus disaggregated by the state of the document

State	Documents	Sentence pairs	Tokens (eu/es)
Finished	1,774	23,695	198,503/236,462
Unfinished	179	2,742	19,569/23,034

TABLE 7.2

Statistics of Itzulbide bilingual corpus disaggregated by document type

Type	Documents	Sentence pairs	Tokens (eu/es)
Hospitalization report	24	625	4,989/5,494
Progress report	1,333	15,069	110,699/127,036
Discharge report	260	4,424	29,660/33,174
Informative permission	139	3,006	42,193/55,639
Others	197	3,313	30,531/38,153

including unfinished documents that could have been written only in one language; while the number of sentences and tokens correspond to the bilingual corpus obtained after automatic extraction and manual alignment.

Both in the following Tables and main text, we use 'eu' and 'es' abbreviations to refer to Basque and Spanish languages respectively, corresponding to the standard ISO 639-2 codes commonly used in natural language processing.

7.3.2 OTHER BILINGUAL CORPORA FROM THE HEALTH DOMAIN

Given the need for big quantities of in-domain data for training state-of-the-art Neural Machine Translation (NMT) systems, our main priority has been to collect as much as possible bilingual corpora from the health domain. To this end, apart from carefully preprocessing the 26,437 bilingual sentences from Itzulbide described in section 3.1, we have compiled 541 more bilingual sentences extracted from 17 clinical cases written in the Basurto hospital (Magnini et al. 2020). These clinical cases are available on the web[7], and specifically, the ones we used were written during 2014 and 2015. Note that these documents are categorized by different specialties than the ones used in the Itzulbide corpus, so these 541 sentences will not be tagged by specialty for the experiment described at the end of this section.

In addition, we have downloaded documents for professional health workers published by Osakidetza[8]. From all the documents available on that website, we omitted the administrative ones (in Spanish: *'Planes y programas anuales y plurianuales'* and *'Memorias Osakidetza'*) and only made use of the documents that were available in both Basque and Spanish in the date of download (October 1, 2020).

TABLE 7.3

Statistics of Itzulbide bilingual corpus disaggregated by specialty

Specialty	Documents	Sentence pairs	Tokens (eu/es)
Oral and maxillofacial surgery	13	86	782/835
Oral and maxillary surgery	1	22	323/453
Anesthesia and resuscitation	22	124	812/947
Respiratory system	51	2,360	16,893/17,620
Internal medicine	182	4,392	31,201/35,922
Digestive system	39	1,183	8,349/9,836
Short stay psychiatry	7	120	1,381/1,507
Out-of-hospital emergencies	4	2	9/36
Nursing	156	1,101	10,537/13,413
Diagnostic radiology	40	240	1,738/2,135
Rehabilitation	10	53	321/338
Ongoing care	51	413	3,827/4,276
Home hospitalization	76	642	4,077/4,685
Unknown	104	2,413	29,594/38,624
Family medicine	251	2,483	17,198/19,023
Pharmacy	50	795	8,534/11,024
Gynecology and obstetrics	4	53	315/348
Cardiology	1	25	213/280
General surgery	2	15	128/134
Health management unit	1	22	467/639
Emergency department	226	2,940	20,003/22,043
Intensive care medicine	33	643	4,811/5,674
Otorhinolaryngology	52	1,019	6,260/7,319
Pediatrics	74	374	2,804/3,267
Preventive medicine	1	23	108/123
Psychiatry	172	746	6,619/8,165
Trauma	104	1,219	9,054/10,950
Urology	125	2,751	29,500/37,306
Palliative care	97	156	1,915/2,278
Palliative care unit	3	21	286/275
Management	1	1	13/21

These documents were converted from pdf to text, and later sentence segmentation was performed for the documents in each language, using an in-house program that was also used for the Itzulbide bilingual corpus. Finally, sentences in both languages were manually aligned; and for making the most of the scarce in-domain bilingual corpora, in case one sentence in one language corresponded to many sentences in the other language, these many sentences were joined using ';' as a separator. In this way, we obtained 22,051 more parallel sentences from the health domain.

7.3.3 Clinical Terminologies Used as Sentences

As additional bilingual in-domain data, we have extracted clinical terminologies from different sources. Even if these clinical terms are not actually sentences, considering the rich vocabulary of health domain and the low resources for eu/es language pair, we think that they can be useful for improving the coverage of the NMT system, helping to translate clinical terms that probably do not appear in the few available bilingual sentences. Moreover, since the sentences we want to translate are usually short and often omit verbs, we consider that using clinical terms formed by a few tokens as sentences will not have any negative effect on the final performance of our system.

Most of the clinical terminologies we have used come from the automatic translation into Basque of the Systematized Nomenclature of Medicine – Clinical Terms (SNOMED CT), which is described as the most comprehensive clinical terminology collection in the world (Perez-de-Viñaspre 2017). The system automatically creates Basque terms corresponding to the original English terms by combining the use of dictionaries, transliteration tools, and rule-based systems, and for training our MT models we used all the terms containing up to 11 tokens (896,898 in total). As these terms are automatically created in Basque, there is often more than one term in Basque for each concept, which we think can be especially useful for translating in the eu-es direction.

Another clinical terminology collection we have used is the International Statistical Classification of Diseases and Related Health Problems 10th Revision (ICD-10). In our case, we have used the manual translation of the concept descriptions in Basque as provided by the organizers of WMT Biomedical shared task (Bawden et al. 2020), obtaining the corresponding descriptions in Spanish directly from ICD-10. From this process we compiled 27,696 additional segments.

Finally, as an initial step for making the systems ready to translate COVID-19 related terms, we have compiled a few dictionaries from an interim release of SNOMED CT[9], consisting of 84 terms, having the English terms translated into Basque by a translator of Osakidetza. Additionally, we compiled 126 COVID-19 related terms compiled by Elhuyar foundation, including all the terms published until June 18, 2020.

7.3.4 Bilingual Out-of-Domain Corpora

When looking for out-of-domain bilingual corpora that could be useful for our task of translating clinical notes, we have looked for a balance between compiling as much data as possible (given the lack of resources) and granting a minimum quality (given the desired high accuracy for a sensitive domain like the health domain). In this sense, we have chosen two corpora that have been professionally translated and, whether have been previously tested on eu/es NMT, or they have similar characteristics to the clinical texts we want to translate.

The first and bigger out-of-domain corpus was originally composed by 4.5 M sentences, being half of them a repetition of a corpus from the news domain (Etchegoyhen and Gete 2020), and the other half coming from diverse sources such as administrative texts, web-crawling, and specialized magazines. This corpus has been previously used for MT of clinical texts between Basque and Spanish (Soto et al. 2019a, 2019b, 2020). For reducing the noise introduced by out-of-domain sentences, we have applied a language identification tool[10] to exclude sentences where most of the terms are named entities like locations or person names. This way, the vocabulary of the out-of-domain corpus is reduced, so a bigger part of the limited vocabulary of the NMT system can be used for translating health domain terms. By removing the sentences that are classified as another language in each of the eu and es sides of the corpus, we filtered 3,703,757 sentences from the original 4.5 M sentences.

The other out-of-domain corpus we use is HAC (Sarasola et al. 2015), compiled by OPUS (Tiedemann 2012) and formed by 566,738 sentences coming from the translation of literary books. Even if the domain is very different from the health domain, we chose this corpus for being translated by professional translators and having similar average sentence length to our clinical domain corpus. On the contrary, we discard other publicly available corpora like OpenSubtitles or GNOME for not having the desired translation quality.

Regarding the preprocessing of the out-of-domain corpora, given that the maximum number of tokens in the clinical domain corpus is 98, we removed all sentences longer than 100 tokens using Moses tools[11]. We also tried to remove sentences shorter than three tokens or using punctuation normalization tools, but both experiments decreased the performance of our NMT system as shown in preliminary results. All of the corpora, including the monolingual ones presented in the next section, were tokenized and truecased using Moses tools[12], being the Truecase model learned on the bigger out-of-domain corpus with its original 4.5 M sentences.

Table 7.4 sums up the statistics of the bilingual corpora described in sections 7.3.1–7.4.

7.3.5 SPANISH MONOLINGUAL CORPORA FROM THE HEALTH DOMAIN

For the eu-es translation direction, we leveraged the Electronic Health Records (EHR) in Spanish already compiled from Osakidetza in previous research projects.

TABLE 7.4

Description and statistics of the diverse bilingual corpora

Corpus	Sentence pairs	Tokens (eu/es)
Itzulbide bilingual corpus	26,437	218,072/259,496
bilingual sentences from Basurto hospital	541	5,254/5,185
Osakidetza's publications for health workers	22,051	299,203/350,361
SNOMED CT clinical terms	896,898	3,074,750/5,309,227
ICD-10 terms	27,696	229,248/175,627
COVID-19 related terms from SNOMED CT	84	579/729
COVID-19 related terms from Elhuyar	126	263/243
out-of-domain corpus (news and others)	3,703,757	66,284,429/95,714,868
HAC corpus (literary)	566,738	8,861,175/10,956,345

For the experiments performed in this work, we have used discharge reports from two hospitals: Galdakao-Usansolo and Basurto. The EHRs from Galdakao-Usansolo hospital consist of 142,154 documents compiled from 2008 to 2012, while the discharge reports from Basurto hospital sum up to 57,569 documents written in 2014. After performing sentence segmentation using the same preprocessing applied to the bilingual in-domain corpora, removing repeated sentences in the corpus from each hospital, and deleting sentences only containing the document ID and/or date, we obtained 1,921,672 sentences from Galdakao-Usansolo and 905,893 from Basurto.

Due to privacy issues, these corpora cannot be made publicly available. The documents were given to us without any personally identifiable information, and before using the corpus from each hospital, it was further de-identified by means of shuffling the sentences. Only authors who had previously signed a non-disclosure commitment had access to them.

Table 7.5 shows the statistics of the Spanish monolingual corpora from the health domain described in this section. In this table, the number of documents corresponds to the total number of compiled documents; while the number of sentences and tokens correspond to the corpus obtained after filtering and preprocessing.

7.3.6 System Training and Evaluation

Being NMT the state-of-the-art method for MT, and based on previous work on translation between Basque and Spanish (Etchegoyhen et al. 2018), we use NMT for training our systems. More specifically, taking into account previous work on the translation of clinical texts from Basque to Spanish (Soto et al. 2019b), we choose the Transformer (Vaswani et al. 2017) architecture. From the different available implementations, we decided to use Fairseq (Ott et al. 2019), for being the one mostly used in recent MT shared tasks (Bawden et al. 2020).

Considering the rich morphology of Basque language and the rich terminology of health domain texts, we use Byte Pair Encoding (BPE) (Sennrich et al. 2015) with 90,000 merge operations for subword segmentation. Additionally, we try BPE-dropout (Provilkov et al. 2020) with 0.1 probability for preprocessing our training corpora, whether applied in both sides of the training corpus or only in the source side. We believe this regularization technique can be especially useful in our rich

---

TABLE 7.5

Statistics of the Spanish monolingual corpora from the health domain

Hospital	Documents	Sentences	Tokens
Galdakao-Usansolo	142,154	1,921,672	32,084,578
Basurto	57,569	905,893	11,812,057

vocabulary setting, and can improve the robustness of the system against the usual typos appearing in EHRs.

As a general method for domain adaptation, we use regular fine-tuning, using the corpora most similar to our evaluation data for fine-tuning, and the rest for pre-training. In this sense, after extracting the corresponding evaluation corpus from the Itzulbide bilingual corpus (1,000 sentences for validation and 1,000 for testing), we use the remaining 24,437 sentences for fine-tuning. In addition, we also use the 541 bilingual sentences from Basurto hospital described in the first paragraph of section 3.2 for fine-tuning. The rest of the bilingual corpora presented in Table 7.4 is used for pretraining.

For both translation directions, we pretrained the systems for 20 epochs when using BPE and for 50 epochs when applying BPE-dropout. In each case, we fine-tuned the systems for the same number of epochs, calculated the BLEU (Papineni et al. 2002) scores on the validation set in each of the models saved after every epoch, and finally calculated the BLEU score on the test set with the model that obtained the highest BLEU on the validation set.

We set two different configurations for evaluating our es-eu models: in the first one, apart from testing the systems preprocessed with different subword segmentation methods, we wanted to evaluate how our system would work in a specialty for which we have no training data. For simulating this scenario, we extracted all the finished sentences from trauma specialty before extracting the evaluation corpus, and used all of these sentences from trauma as a second test set. The sentences corresponding to trauma specialty coming from documents marked as unfinished or from the 541 bilingual sentences from Basurto hospital were not used for training in this scenario. We chose this specialty for having around 1,000 finished sentences in the Itzulbide bilingual corpus and having a distinct terminology from the other specialties. Thus, we will have six different results from this configuration, corresponding to the use of BPE, BPE-dropout and BPE-dropout applied only in the source language for subword segmentation; and evaluated both in the trauma test set and in the test set extracted from the remaining specialties.

In the second scenario, our aim was to develop a system that matched the requirements of Osakidetza as closely as possible, so we used sentences from all the specialties for both training and evaluation, and selected only discharge reports and progress reports for evaluation, using the sentences coming from other document types only for training. For training this system we used the subword segmentation method that obtained the best results in the previous scenario; and additionally, we tried adding tags for identifying the specialty of the sentences from Itzulbide used for fine-tuning. For doing this, we defined an acronym of three characters for each specialty listed in Table 7.3, and include it between '<' and '>' marks (for instance, '<DIG>' for digestive system). Then, for each Basque sentence in the Itzulbide bilingual corpus, we inserted the corresponding specialty tag at the beginning of the sentence, separated by a blank space. Specifically, we inserted the tag after applying subword segmentation, so the tags did not influence the learned BPE model. After evaluating the systems with and without using these tags, we used the best performing system resulting from this process for translating the available EHRs in

Spanish into Basque. The translation was done using unrestricted sampling (Edunov et al. 2018) as decoding method[13], with a buffer size of 1,000 and a batch size of 50.

The created synthetic corpus was added to the training corpora used in the es-eu direction for training the systems in the eu-es direction. We conducted two experiments using this synthetic corpus: in the first one we added the pseudo-parallel corpus created via back-translation (Sennrich et al. 2016), and in the second one we further added the monolingual corpus as both source and target corpus, using the technique known as copying (Currey et al. 2017). In both experiments, we applied BPE-dropout and pretrained/fine-tuned the systems for 50 epochs.

7.4 RESULTS AND DISCUSSION

7.4.1 Spanish to Basque Translation Direction

First, we compare the results in the es-eu translation direction when using BPE, BPE-dropout, or BPE-dropout applied in the source side only. Table 7.6 presents the BLEU scores obtained with these subword segmentation methods, evaluated in the trauma test set and in the general test set formed by sentences from other specialties.

We observe that BPE-dropout obtains the best results in both test sets, so we use this subword segmentation method for further experiments. Regarding the big difference between the BLEU scores obtained in the general test set and the trauma test set, one would think that the performance in the held-out specialty is very low, but after looking at the generated translations we see a different picture: most of the differences between the human translations and the machine translations do not correspond to errors of the automatic translations, but to errors on the manual translations. These differences can be classified into diverse categories, like typos ('*egunena*' instead of '*egunean*', meaning 'in the day'), orthographic errors ('*protezi*' instead of '*protesi*', meaning 'prosthesis'), or even the use of different dialects of Basque ('*ondo*' instead of '*ongi*', meaning 'good'). In the analyzed 100 sentences we did not observe any error that could be attributed to the domain shift; however, a thorough human evaluation should be performed by healthcare workers to study the effect of using our MT tool to translate sentences from a specialty not seen in the training corpus.

TABLE 7.6

BLEU scores in the es-eu translation direction using different subword segmentation methods (best results in each test set are marked in bold)

Subword segmentation method	BLEU (general test set)	BLEU (trauma test set)
BPE	36.24	19.28
BPE-dropout	**37.42**	**19.52**
BPE-dropout (source side only)	37.06	18.57

In the second scenario, we tested the effect of adding tags to the diverse specialties included in the bilingual corpus from Itzulbide. As stated in the previous section, in this setting we evaluate only onsentences coming from progress reports and discharge reports. Table 7.7 shows the BLEU scores obtained with and without specialty tags (for completion, the BLEU score of the system before fine-tuning is 18.28).

We observe that the system without tags performs better, so we use this model for translating the Spanish monolingual corpora from the health domain and do not use tags for the following experiments in the eu-es translation direction.

7.4.2 BASQUE TO SPANISH TRANSLATION DIRECTION

For the eu-es translation direction, we perform two experiments including the Spanish monolingual corpora from the health domain in two different ways: one including the pseudo-parallel corpus resulting from back-translation (Sennrich et al. 2016), and the second one applying also the technique known as copying (Currey et al. 2017). Table 7.8 presents the BLEU scores for these settings. For a better analysis, we include the results before and after fine-tuning, and also show the size of the pretraining corpus in each configuration (the fine-tuning corpus is formed by 24,978 sentences).

Contrary to previous work that did not use bilingual clinical domain data (Soto et al. 2019a), we observe that in our case the technique of copying does not improve the results, so we present the system using only back-translation as the best performing system.

TABLE 7.7

BLEU scores in the es-eu translation direction with and without specialty tags (best result is marked in bold)

Use of tags	BLEU (progress reports and discharge reports)
without tags	**31.97**
with tags	30.67

TABLE 7.8

Number of sentences of the pretraining corpus, along with BLEU scores in the eu-es translation direction after back-translation and further copying; before and after fine-tuning (best result in each case is marked in bold)

Method	Pretraining sentences	BLEU (before fine-tuning)	BLEU (after fine-tuning)
+back-translation	7,956,799	38.50	**50.67**
++copying	10,729,257	**38.74**	49.64

7.4.3 ERROR ANALYSIS

In this last subsection, we analyze the best-performing systems in both translation directions, which obtained 31.97 BLEU points in es-eu and 50.67 in eu-es. Before analyzing the generated translations, note that BLEU metric underestimates the translation performance into morphologically rich languages such as Basque; so, even if it is expectable that the eu-es system performs better thanks to the use of Spanish monolingual EHRs, part of the big gap between the BLEU scores in es-eu and eu-es can also be attributed to this fact.

As a first simple analysis of the generated translations, we checked if numbers were correctly translated, given the special relevance it could have to incorrectly translate a result of a medical test, a prescribed dose, an appointment date, hour, etc. With this aim, we manually checked how numbers were translated in the first 100 sentences of the test set in both translation directions, and confirmed that all numbers were translated correctly.

Secondly, given the lack of standardized clinical terminology in Basque, we wanted to analyze how clinical terms were translated in the es-eu direction. For doing this semi-automatically, we used the 896,898 bilingual terms coming from the automatic translation of SNOMED CT into Basque as a reference, looking for their appearance in both sides of our test set translated by the best performing system. Since the SNOMED CT terms in Basque were automatically created, this analysis is conditioned by the way this term creation was done. That is why, when looking for these terms in the test set, we observed that all the clinical terms identified in both Basque and Spanish corresponded to terms that were identically written or transliterations, as this was one of the methods used for automatically creating the Basque terms. Specifically, we detected 44 bilingual terms in 37 sentences out of the 1,000 sentences of the test set, and observed that all of them were translated correctly. The only differences between the terms included in the generated translations and the automatically created SNOMED CT terms in Basque corresponded to one appearance of the acronym '*EKG*' instead of the extended '*elektrokardiograma*' (meaning 'electrocardiogram') and three appearances of '*arnas-hestu*' or the shrinked '*arnasestu*' instead of the recommended '*disnea*' (meaning 'dyspnea'). In the latter case, we can consider that the generated translations ('*arnas-hestu*' / '*arnasestu*') are more informal than the more technical '*disnea*', but we can anyways accept them as correct translations for being similar to the 'breathless' term found in SNOMED CT as a synonym of 'dyspnea' ('*arnas*' meaning 'breath' and '*hestu*' meaning 'tight').

7.5 CONCLUSIONS AND FUTURE WORK

Analyzing the automatic evaluation results shown in the previous section, we can conclude that the developed systems obtain good results in both translation directions, especially in the eu-es direction thanks to the big number of EHRs available in Spanish (around 2.8 M sentences). We also observe that using a bilingual in-domain corpus for fine-tuning, even if limited to 24,978 sentences, greatly improves the final performance of the systems, boosting 14 BLEU points when translating from Spanish into Basque, and 12 from Basque into Spanish. This validates our effort of compiling the bilingual clinical domain corpus as a necessary first step for developing a domain-adapted MT model.

Regarding the translation of clinical terminologies, we have observed that our systems always generate correct terms, and most of the time use the terms coming from the previous automatic translation into Basque of SNOMED CT. However, we believe that it would be helpful for future work to have a standardized clinical terminology in Basque, so we could properly evaluate the MT of clinical terms.

Regarding data preprocessing, we have proved that the use of a regularization technique like BPE-dropout improves the results in a domain where typos, mis-spellings, etc. are usual, giving us an extra BLEU point in the es-eu translation direction. Additionally, we have tried using tags for identifying the specialties in the clinical domain bilingual corpus, but saw no gains from including them as a way to guide the NMT system, probably due to the limited number of training examples for some of the specialties.

Overall, even if the automatic evaluation scores and error analysis show promising results, a human evaluation should be performed before implementing these systems for translating clinical texts in a real-life scenario. It would also be interesting to extend this human evaluation for testing the performance of the models in a held-out specialty, as we did in this work by automatic means. We leave this human evaluation as future work.

Furthermore, we plan to keep compiling bilingual/monolingual clinical domain corpora to continuously improve the performance of our systems, focusing on the eu-es translation direction. With this aim, we plan to try diverse strategies for back-translation (Caswell et al. 2019; Graça et al. 2019; Hu et al. 2019) and apply data selection methods to the back-translated data (Soto et al. 2020).

Up to now, we have developed a useful MT tool, obtaining 50.67 BLEU points, that could help Spanish monolingual speakers understand the clinical notes written in Basque. However, in case this system is implemented in a real-world scenario, given the critical effects it could have a decision based on an incorrect translation, we recommend that a bilingual speaker reviews the generated translations and corrects them if necessary. In any case, the deployment of this model can encourage Basque-speaking healthcare workers to write their reports in Basque, which is an objective of the Itzulbide project that has already started to be fulfilled with the 26,437 sentences used for building this initial MT tool. Furthermore, these clinical domain corpora can also be helpful for other NLP tasks.

NOTES

1 https://en.wikipedia.org/wiki/Basque_Country_(greater_region)
2 https://en.wikipedia.org/wiki/Basque_language
3 This paragraph has been adapted from Perez-de-Viñaspre (2017).
4 https://www.osakidetza.euskadi.eus/transparencia-buen-gobierno/-/organigramas-de-organizaciones-de-servicios-de-osakidetza/
5 It is well known that the type of language used by healthcare workers is not specially correct. This is mainly due to the fact that they often write quickly, which leads them to

make typing mistakes, write sentences that do not follow grammatical rules, and use made up expressions or acronyms.

6 We are aware that the fact that most of the doctors and nurses write their health records initially in Spanish, and later translate them into Basque, makes the further evaluation in the Basque-to-Spanish direction suboptimal, but we are constrained by the fact that currently, healthcare workers are forced to write the documents in Spanish; and given the lack of Basque standardized clinical terminology, writing them initially in Basque would require more effort from the physicians while they are dealing with their patients.

7 https://github.com/hltfbk/E3C-Corpus

8 https://www.osakidetza.euskadi.eus/profesionales/-/publicaciones-profesionales/

9 https://www.snomed.org/news-and-events/articles/march-2020-interim-snomedct-release-COVID-19

10 https://github.com/saffsd/langid.py

11 https://github.com/moses-smt/mosesdecoder/blob/master/scripts/training/clean-corpus-n.perl

12 https://github.com/moses-smt/mosesdecoder/blob/master/scripts/tokenizer/tokenizer.perl and https://github.com/moses-smt/mosesdecoder/blob/master/scripts/recaser/truecase.perl respectively.

13 For details about how to implement unrestricted sampling in fairseq, see: https://github.com/pytorch/fairseq/issues/308

REFERENCES

Bawden, R., G. Di Nunzio, and C. Grozea et al. 2020. Findings of the WMT 2020 biomedical translation shared task: Basque, Italian and Russian as new additional languages. In *5th Conference on Machine Translation (WMT2020)*. Association for Computational Linguistics. Online, European Language Resources Association, Marseille, France.

Caswell, I., C. Chelba, and D. Grangier. 2019. Tagged back-translation. In *Proceedings of the Fourth Conference on Machine Translation (WMT2019)*. Association for Computational Linguistics. Florence, Italy.

Currey, A., A. V. Miceli-Barone, and K. Heafield. 2017. Copied monolingual data improves low-resource neural machine translation. In *Proceedings of the Second Conference on Machine Translation (WMT2017)*. Association for Computational Linguistics. Copenhagen, Denmark.

Desjardins, L. 2003. *La santé des francophones du Nouveau-Brunswick*. Petit-Rocher: Société des Acadiens et des Acadiennes du Nouveau-Brunswick.

Etchegoyhen, T., and H. Gete. 2020. Handle with care: A case study in comparable corpora exploitation for neural machine translation. In *Proceedings of the 12th Language Resources and Evaluation Conference (LREC 2020)*. European Language Resources Association. Marseille, France.

Edunov, S., M. Ott, M. Auli, and D. Grangier. 2018. Understanding back-translation at scale. arXiv preprint arXiv:1808.09381

Etchegoyhen, T., E. Martínez, and A. Azpeitia et al. 2018. Neural machine translation of Basque. In *Proceedings of the 21st Annual Conference of the European Association for Machine Translation (EAMT 2018)*. European Association for Machine Translation. Alacant, Spain.

European Observatory on Health Care Systems. 1999. *Health Systems in Transition*. Luxembourg: Health system review.

Gerkens, S., and S. Merkur. 2010. Belgium: Health system review. *Health Systems in Transition* 12, no. 5: 1–266.

Graça, M., Y. Kim, J. Schamper, S. Khadivi, and H. Ney. 2019. Generalizing back-translation in neural machine translation. In *Proceedings of the Fourth Conference on Machine Translation (WMT2019)*. Association for Computational Linguistics. Florence, Italy.

Hu, J., M. Xia, G. Neubig, and J. Carbonell. 2019. Domain adaptation of neural machine translation by lexicon induction. In *Proceedings of the 57th Annual Conference of the Association for Computational Linguistics (ACL 2019)*. Association for Computational Linguistics. Florence, Italy.

Joanes Etxeberri Saria V. Edizioa. 2014. *Donostia unibertsitate ospitaleko alta-txostenak*. Komunikazio Unitatea: Donostiako Unibertsitate Ospitalea.

Magnini, B., B. Altuna, A. Lavelli, M. Speranza, and R. Zanoli. 2020. The E3C project: Collection and annotation of a multilingual corpus of clinical cases. In *Proceedings of the Seventh Italian Conference on Computational Linguistics (CLiC-it 2020)*. Accademia University Press. Online.

Ott, M., S. Edunov, and A. Baevski et al. 2019. Fairseq: A fast, extensible toolkit for sequence modeling. In *Proceedings of NAACL-HLT 2019: Demonstrations*. Association for Computational Linguistics. Minneapolis, Minnesota, USA.

Papineni, K., S. Roukos, T. Ward, and W. Zhu. 2002. BLEU: A method for automatic evaluation of machine translation. In *Proceedings of the 40th Annual Meeting of the Association for Computational Linguistics (ACL 2002)*. Association for Computational Linguistics. Philadelphia, Pennsylvania, USA.

Perez-de-Viñaspre, O. 2017. Automatic medical term generation for a low-resource language: Translation of SNOMED CT into Basque. PhD thesis, Donostia, Spain. University of the Basque Country.

Provilkov, I., D. Emelianenko, and E. Voita. 2020. BPE-dropout: Simple and effective subword regularization. In *Proceedings of the 58th Annual Meeting of the Association for Computational Linguistics (ACL 2020)*. Association for Computational Linguistics. Online.

Sarasola, I., P. Salaburu, and J. Landa. 2015. *Hizkuntzen Arteko Corpusa (HAC)*. Bilbao, Spain. University of the Basque Country UPV/EHU (Euskara Institutua).

Sennrich, R., B. Haddow, and A. Birch. 2015. Neural machine translation of rare words with subword units. arXiv preprint arXiv:1508.07909

Sennrich, R., B. Haddow, and A. Birch. 2016. Improving neural machine translation models with monolingual data. In *Proceedings of the 54th Annual Meeting of the Association for Computational Linguistics (ACL 2016)*. Association for Computational Linguistics. Berlin, Germany.

Soto, X., O. Perez-de-Viñaspre, G. Labaka, and M. Oronoz, (2019a). Neural machine translation of clinical texts between long distance languages. *Journal of the American Medical Informatics Association* 26, no. 12: 1478–1487.

Soto, X., O. Perez-de-Viñaspre, M. Oronoz, and G. Labaka. (2019b). Leveraging SNOMED CT terms and relations for machine translation of clinical texts from Basque to Spanish. In *Proceedings of the Second Workshop on Multilingualism at the Intersection of Knowledge Bases and Machine Translation*. European Association for Machine Translation. Dublin, Ireland.

Soto, X., D. Shterionov, A. Poncelas, and A. Way. 2020. Selecting backtranslated data from multiple sources for improved neural machine translation. In *Proceedings of the 58th Annual Conference of the Association for Computational Linguistics (ACL 2020)*. Association for Computational Linguistics. Online.

Tiedemann, J. 2012. Parallel Data, Tools and Interfaces in OPUS. In *Proceedings of the 8th International Conference on Language Resources and Evaluation (LREC 2012)*. European Language Resources Association. Istanbul, Turkey.

Vaswani, A., N. Shazeer, and N. Parmar et al. 2017. Attention is all you need. In *Advances in Neural Information Processing Systems (NeurIPS 2017)*. Curran Associates, Inc. Long Beach, California, USA.

8 Clinical NLP for Drug Safety

Bhasha Khose and Yashodhara V. Haribhakta
Department of Engineering & Information Technology,
College of Engineering, Pune, Maharashtra, India

CONTENTS

8.1   ADR MONITORING SYSTEMS

Nationwide pharmacovigilance systems are deployed across the world to study and monitor the unwanted long-term or short-term impact of pharmaceuticals. A nation's healthcare department specifically has a drug administration and regulation agency that maintains Adverse Drug Reaction (ADR) database for drug monitoring and management. The nations coordinate with the international drug monitoring systems governed by the World Health Organization (WHO). WHO has the Collaborating Center for International Drug Monitoring, Uppsala Monitoring Center (UMC), in Sweden. The ADRs database in Uppsala has over 3 million reports of suspected ADRs as of now.

The Legatum Prosperity Index 2020 shows that top six regions from a health perspective are North America, Western Europe, Eastern Europe, Latin America, the Caribbean, and Asia Pacific. One of the criteria to evaluate these health system rankings is robust healthcare system and preventative interventions (https://www.prosperity.com/download_file/view/4193/2025).

North American countries like the United States have the Food and Drug Administration (FDA) agency within the Department of Health and Human Services. As part of the U.S. FDA, the Center for Drug Evaluation and Research (CDER) regulates over-the-counter and prescription drugs, including biological therapeutics and generic drugs. The leading consumer watchdog in this system is FDA's Center for Drug Evaluation and Research (CDER). FDA Regulates $1 trillion worth of products a year. The U.S. FDA launched MedWatch – The FDA Safety Information and Adverse Event Reporting Program (US Food and Drug Administration 2016). Health Canada is a federal regulatory institution in Canada that mandates reporting serious ADRs and MDIs by hospitals. Hospitals are required to report serious ADRs and MDIs for prescription and non-prescription medications, biologic drugs, radiopharmaceuticals, disinfectants, medical devices, and drug needs for emergency public health purposes (Government of Canada 2020).

France in Western Europe has the French Pharmacovigilance System to maintain the French Pharmacovigilance Database (FPD), which records spontaneous reporting of ADRs; reporting "serious" or "unlabelled" ADRs to the French Regional Centers has been mandatory for any drug prescriber, physician, dentist, or midwife in France since 1995 (Tavassoli et al. 2009). The European Medicines Agency (EMA) is a decentralized agency of the European Union (EU) responsible for the scientific evaluation, supervision, and safety monitoring of medicines in the EU. EudraVigilance is the system for managing and analyzing information on suspected adverse reactions to medicines that have been authorized or being studied in clinical trials in the European Economic Area (EEA) (European Medicine Agency).

China has China Food and Drug Administration (CFDA) to ensure the safety and effectiveness of medication use. The China Adverse Drug Reaction Monitoring System (CADRMS) is an online spontaneous reporting system that connects the four-level pharmacovigilance network (national, provincial, municipal, and county) (Zhang et al. 2014).

Drug safety plays a critical role in healthcare and drug development lifecycle. Drug safety systems are valuable mechanism to approve new medication or to withdraw existing medicines from the market. Every drug launched in the market for public usage must acquire drug regulatory authorities' approval before entering the market. Drug regulatory authorities monitor and assess the drug for any potential drug-related problem in the pre-market (clinical trials) and post-market (reporting) phase before approving the drug for public usage. The efficacy of the national drug safety system depends on the awareness and commitment of health care workers and patients towards ADR reporting. ADR reporting by the healthcare workers is the key for drug authorities to intervene, assess, analyze, and recommend required actions to implement drug safety.

8.2   ADR REPORTING PROCESS

ADR reporting and monitoring process can be generalized and divided in to five key phases. These key phases are captured in Figure 8.1. The mentioned process starts with reporting; patients and healthcare professionals (HCPs) report ADRs to

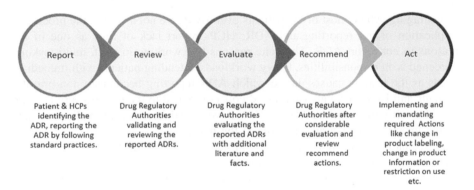

Report	Review	Evaluate	Recommend	Act
Patient & HCPs identifying the ADR, reporting the ADR by following standard practices.	Drug Regulatory Authorities validating and reviewing the reported ADRs.	Drug Regulatory Authorities evaluating the reported ADRs with additional literature and facts.	Drug Regulatory Authorities after considerable evaluation and review recommend actions.	Implementing and mandating required Actions like change in product labeling, change in product information or restriction on use etc.

FIGURE 8.1 Common drug safety process.

authorities using nationwide drug regulatory systems. These systems have pre-defined processes to capture the details of patient, treatment, and ADR. Generally, there is an online and offline mode to capture and report the ADR. Different ADR forms differentiated using colors, form numbers based on the ADR reporter's profile [Patient or HCP]. Serious events reporting also has special focus. There are separate categorization and processes for evaluating serious drug reactions.

Reported ADRs are further analyzed, categorized, and prioritized based on the impact and availability of evidence, which is the second step. Patient history, demographics, and the criticality of the ADR is reviewed to validate the report. Though it is easier to decide the validity of the ADR reported, in some scenarios, an in-depth and thorough analysis is required to establish the ADR. Further analysis involves validating the clinical relevance, exposure, temporal association, biological plausibility, de-challenge and rechallenge, severity of reaction, and outcome, which are important considerations and the basis of the suspected reaction and previous awareness of the association (European Medicine Agency). Available biomedical literature and information sources are used to find the relevant associations.

After a stringent review and evaluation of the reported ADRs, confirmed ADRs are further assessed by Drug Authorities, and recommendations are proposed for safe use of the drug. Based on the seriousness of the ADR various recommendations can be proposed, for example: updating the product label for including or modifying the existing information, awareness and educational programs for patient and HCPs for safe use of drug, providing safety updates, or restricting the use of the medicine.

Recommended actions are then implemented by the drug manufactures. Recommendations are communicated and announced by the drug authorities. Drug authorities may choose various mode of communication to increase awareness in patients and HCPs based on the severity of ADR.

Healthcare systems in the world facilitate reporting ADRs, however, looking at the historical records of ADR reporting across the globe, there is huge gap and reluctance to report the ADRs.

There are surveys conducted on patients to know if they are aware of what is ADR, how to detect ADR and how it can be reported, and the results are quite discouraging, where most of the patients even HCPs are not aware of the process or implication of not reporting and ADR. HCPs report lack of time as one of the reasons to not report ADR. They cite reasons like working in shift, multi-tasking, burdened with responsibilities, heavy workloads, attending patients with immediate needs as the leading causes due to which ADR reporting becomes a non-priority. Patients and HCPs have also expressed concern related to the patients' privacy and confidentiality for reporting ADRs. HCPs have shared fear of legal implications and blame for mis-administering the drug at the first place.

In scenarios where HCPs show commitment to report ADR, they find the process unnecessarily complex, blame document work, lengthy procedures, and lack of systematic reporting process to capture the ADR-related information. Historical records also show that known ADRs get over reported and duplicated, where unknown ADRs are under reported by the HCPs as they have less apprehension to known ADRs.

All these challenges warrant the need for revisiting the ADR reporting process. There is need to simplify the process, making it more accessible to patients and HCPs. Key contributors, stakeholders, and users of the process have to be educated and made aware of the implications and need of ADR reporting. Regulatory authorities must address concerns and make population aware of the privacy, confidentiality and legal implications and provide required assurance. Use of cutting-edge technologies like mobile apps, games, and rewards should be explored for building an efficient system, reaching to the patients and HCPs, and establishing informal and open channel of communication for sharing and reporting.

8.3   ADR REPORTING AND CLINICAL NLP

Adverse events capture the undesirable side effect caused after consumption of the drug but every adverse event cannot be classified as adverse drug reaction. Adverse event needs to be studied in the context of drug-dosage, drug-frequency, drug-prescription to label it as adverse reaction.

Following are the formal definitions of adverse drug event and adverse drug reaction:

An Adverse Drug Event (ADE) is "an injury resulting from the use of a drug. Under this definition, the term ADE includes harm caused by the drug (Adverse Drug Reactions (ADR) and overdoses) and harm from the use of the drug (including dose reductions and discontinuations of drug therapy). Adverse Drug Events may result from medication errors but most do not". (VA Center of Medication and Safety 2006).

An Adverse Drug Reaction (ADR) is a "response to a drug which is noxious and unintended, and which occurs at doses normally used in man for prophylaxis, diagnosis, or therapy of disease or for the modification of physiologic function". Note that there is a causal link between a drug and an ADR. In sum, an ADR is harm directly caused by the drug at normal doses during normal use. (VA Center of Medication and Safety 2006) (Figure 8.2).

This is one of the reasons there is various levels of mandatory attributes required to report ADR. These mandatory attributes differ from one drug regulatory system

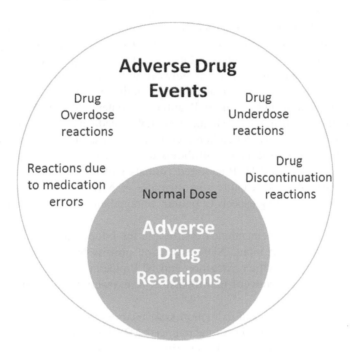

FIGURE 8.2   ADE vs ADR.

TABLE 8.1
ADR key attributes

Information Category	Fields/Attributes
Patient Info	Name, age when reaction occurred
Reporter Info	Details of person reporting the ADR, may be patient or a relative/friend or a HCP
ADR Info	Reaction, medication, onset date, anatomy\location
Additional ADR Info	Intensity\frequency of the reaction

to the other. Table 8.1 captures the higher-level category of information that has to be captured as part of ADR reporting.

ADR reporting is getting automated to increase voluntary reporting and to make the ADR process effortless. Automating processing of clinical reports (Electronic Health Records (EHRs), Electronic Medical Records (EMRs), Patient Health Records (PHRs), Medical Case Reports (MCRs), etc.) and patient narratives obtained from social media, community posts or spontaneous reporting channels for ADR-related information extraction. Clinical NLP methods are employed to first extract candidate phrases, sentences, and span of interests from clinical text and then to extract ADR information as described in Table 8.1.

Primary goal of ADR extraction process from is identifying the biomedical concepts in the clinical text and then to extract drug-ADR pair using various heuristic and deep learning methods. Once the drug-ADR pairs are extracted next step is to Identify the causal relations between drug and event.

ADR extraction process has to adapt based on the input source. Patient authored inputs differs significantly from HCP authored content in various ways. Patient authored content known to have significantly fewer medical words in comparison to HCP authored content, patient authored content lack structured format, mostly will not be capturing key aspects and attributes of reported ADR, and will have high noise content. In contrast HCP authored formats will follow structure, have high density of medical terms, capture key attributes of ADR and will have negligible noise content. Contrasting aspects of patient authored and HCP authored inputs are captured in Table 8.2.

Considering the abovementioned contrast in the information content and structure of the clinical text input. ADR extraction process has to differ and should employ different information extraction and NLP pipeline. Two key parts to ADR extraction process are **concept identification & extraction** task and **relation extraction** task.

Concept identification and extraction task focuses on extracting the candidate tokens and tag them against one of the required ADR info categories as described in Table 8.1. Concept extraction task can utilize various domain agnostic approaches however relation extraction step needs to be performed in the domain context. Researchers have proposed various methods for concept extraction problem. As described in (Song et al. 2015; MacKinlay et al. 2017), pre-annotated database of biomedical concept is created along with predefined rules to extract concepts from input sentences. These dictionary-based approaches are proven to be accurate and robust in case of previously seen input pattern but fails immensely in case of unseen input like variation in spellings, abbreviations, etc. These approaches also require considerable time and efforts of domain expert in order to build this knowledge. Efforts required to create this knowledge can be minimize by employing machine learning algorithm-based approaches similar to the ones proposed by (McCallum et al. 2000; Lafferty et al. 2001; Tsai et al. 2006) that can classify the tokens in particular tags using Naïve Bayes, SVM classifiers, Hidden Markov Models, Maximum Entropy Markov, and CRF models proven to give better results but still poorly performs in case of unseen data. Researchers have proposed deep learning-based approaches that enhance existing Word2Vec, BERT models on domain specific data from Wiki, PubMed, and PMC articles to generate domain specific vector representations (Crichton et al. 2017; Habibi et al. 2017; Wang et al. 2018; Yoon et al. 2019; Lee et al. 2020). Table 8.3 describes the SOT approaches in structured format.

Relation extraction from given clinical text uses contextual information surrounded by the concept mentioned in the sentences. Surrounding verbs and adjectives in the sentences dictate the meaning of the bio-medical concept mentions in the sentence. Researchers employs various heuristic and deep learning-based methods to execute relation extraction task. (Airola et al. 2008; Pyysalo et al. 2008; Kandula and Zeng-Treitler 2010; Kang et al. 2014) Co-occurrence or patterns

TABLE 8.2
Patient vs HCP authored input

	Clinical Study/ Clinical Case Reports	EHRs/EMRs/ PMRs	Spontaneous Reporting	Patient Social Media/ Community Post
Input/ Information Source	Clinical reports capture the case data related to disease, drug, ADRs, methods used, etc. Used to submit case details\trial details to Public health entities.	Patient specific long-term records capture patient's health across time.	Patient/HCP calling the authorities to report the ADR.	Patient sharing details related to his/her medical condition on social media post or in the health community portal.
Author	HCPs	HCPs	Patient/Reporter/ HCPs	Patient
Mode	Active	Active	Passive	Passive
Structure	Well Structured: Clinical reports normally follow a structured pattern and standards.	Semi-Structured: Clinical narrative written in the patient health record is unstructured.	Unstructured: HCPs\Patient will use free form to explain the side effect and there will be no structure to the input.	Free Form: Patient authored text will use free form to describe the medical condition and side effect.
Density of Medical Terms	High: Use standard medical terms to describe the drug, medical conditions and ADRs.	High: Use standard medical terms to describe the drug, medical conditions and ADRs.	Medium: Use Common language words to describe the medical condition. No or less use of standard medical terminology.	Low: Use Common language word, idioms, phrases, and there will be no use of medical terms.
Availability of ADR Related Info	High: Direct mention of Drugs, Dosages, Time, and Frequency of dosages.	High: Direct mention of Drugs, Dosages, Time, and Frequency of dosages.	Medium: Semi-Direct mention of Drugs but there is an opportunity where clinical experts can ask for required ADR related information.	Very Low: No direct mention of Drugs, Dosage, Frequency or timings.
Availability Patient	All the required Patient current	All the required Patient current	Patient current details are	Will not have the complete

(Continued)

TABLE 8.2 (Continued)
Patient vs HCP authored input

	Clinical Study/ Clinical Case Reports	EHRs/EMRs/ PMRs	Spontaneous Reporting	Patient Social Media/ Community Post
Demographics & Medical History	and medical history captured along with Patient demographics.	and medical history captured along with Patient demographics.	captured along with Patient demographics. Patient may not have the history of treatment given.	demographics as well as the medical history data.
Completeness	Captures all the required parameters of ADR.	Mostly Captures all the required parameters of ADR.	May or may not have all the required information. Patient may not have or recall the information. For ex: What drugs he took one day before the side effect occurred.	Mostly will not have all the information required for extracting the ADR.
Noise	Very Low: Clinical report follows standard pattern and its easier to find out ADR related text to process and remove the noise.	Medium: Clinical narratives are HCP biased and subject to different writing styles but due to use of standard way to describe medical conditions and mentions extracting it is possible to extracted candidate text and remove the noise.	High: Transcribes from the call will be used for text mining. Text will be full of extra and irrelevant information. Though it will be difficult to remove the noise but still questions posed by the Clinical expert can be used as guiding factor to extract relevant information.	Very High: Very noisy, context used by the Patient will vary throughout the text ex: Medical condition to load due to House Chores.

(Fundel et al. 2007; Rinaldi et al. 2007; Buyko et al. 2012) specific extraction gives accurate results when input text follows specific template or structure, however, in case of patient authored inputs these co-occurrence or pattern specific approach do not provide accurate results. Training a classifier for extracting relations provides

TABLE 8.3

Concept identification and extraction approaches

Approach	Details	Key Points	Publication References
Dictionary & Grammar Based	Using pre-annotated database to filter sentences containing mention of the biomedical concept. Applying rules/patterns-based logic to extract the actual entities.	• Public pre-annotated English language databases exists for entities like disease, symptoms, chemical, genes etc. • Creating manual annotations are expensive and are created for specific sub-domain. • Face challenges in case of spelling variations, abbreviations, use of hyphens, use of adjectives etc. along with the NER in the input.	(Song et al. 2015; MacKinlay et al. 2017)
Machine Learning (Classifier based)	Uses SVM, Naïve Bayes for classifying the text.	Classify each token in the given text, generating class tag by calculating the score against each class and assign the best.	(Tsai et al. 2006)
Machine Learning (Sequence models)	Uses Hidden Markov Models, Maximum Entropy Markov models and Conditional Random Field based models. Uses IOB formats for the tags.	• Probability based models performs well on the seen data. • CRF based models has shown better performances over HEMM and MEMM models.	(McCallum et al. 2000; Lafferty et al. 2001)
Deep Learning	Training existing word embedding and character embedding generation models like Word2Vec, BERT, on biomedical domain corpora from Wiki, PubMed, PMC etc. to learn domain specific word representations.	• Deep learning-based approach perform better than traditional dictionary matching based approach. • Boundary annotations will be required for learning the start and of entities.	(Habibi et al. 2017; Lee et al. 2020)
Deep Multi-Task Learning (Single Model)	Single model (Bi-LSTM + CRF)is trained to learn word/character embeddings across multiple biomedical domain tasks.	• Handle single Entity type well. • Polysemy is a challenge as only one representation can be generated for the word.	(Wang et al. 2018)

(Continued)

TABLE 8.3 (Continued)

Concept identification and extraction approaches

Approach	Details	Key Points	Publication References
		• Performs well on recall but suffers on the precision values.	
Deep Multi-Task Learning (Multi-Model)	Multiple(Bi-LSTM + CRF) models are used, one model for one biomedical task. Each model specific to a biomedical task is trained on the task specific annotated corpora. Learnings from individual single task models are combined and feed into the target model (Bi-LSTM+ CRF). (Yoon et al. 2019). Another approach used CNN based models. First training a single CNN model on different task specific corpora and then sharing the learnings to the individual task specific CNN models. Another variant also proposed where task-based CNN Model share the learnings. Auxiliary task learnings are shared with main task models. (Crichton et al. 2017).	• Performs better on precision and recall both in compare to SOT. • Attempt to handle the polysemy problem. • Requires large amount of memory and time for training.	(Crichton et al. 2017; Yoon et al. 2019)

better results but requires huge volume of annotated training data (Katrenko and Adriaans 2006). (Huang et al. 2005; Demner-Fushman et al. 2009) proposes hybrid methods combining rules and annotated data to reduce the manual efforts, however, perform poorly in case of unseen patterns. Deep learning-based biomedical word embeddings methods that employs training existing word embedding models like BERT and Word2vec using biomedical domain data to learn domain specific embeddings reduce the expert efforts and perform better in case of unseen data, but requires considerable large unlabeled domain corpus. Table 8.4 consolidates SOT approaches used in this space.

TABLE 8.4

Relation extraction approaches

Approach	Details	Key points	Publication References
Co-occurrence Based	Drug-ADEs pairs are extracted based on how frequently pairs occur in the same sentences across the corpus.	• These methods have shown high Recall and low precision. • Do not consider the semantic or linguistic relation between the pair. • Can-not handle unseen pairs.	(Airola et al. 2008; Pyysalo et al. 2008; Kandula and Zeng-Treitler, 2010; Kang et al. 2014)
Rule Based	Creating linguistic rules based on the context in which drug-ADE pair normally expressed. Rules created taking word feature like POS, suffix, affix, chunks etc.	• Performance is good with high recall but suffer in case of unseen words/ pairs. • Language ambiguity and various ways to express the side effects require huge manual efforts to cover all the possible combinations.	(Fundel et al. 2007; Rinaldi et al. 2007; Buyko et al. 2012)
Machine Learning Based	Training statistical classifier (mostly SVM based) for relation extraction. Uses NLP led feature engineering which is based on syntactic relations.	• Give good results, assuming we have considerable annotated corpus for training the classifier.	(Katrenko and Adriaans 2006)
Hybrid (Rule+ML)	Combining rule-based and machine learning based techniques.	• Require less annotated data and reduce manual efforts for rule building.	(Huang et al. 2005; Demner-Fushman et al. 2009)
Deep Learning	Classifying sentences for relations using BI-LSTM + CRF architecture. Approach to anonymize target entities (ibuprofene->$Drug) using pre-defined tags to learn the relations shows better performance.	• Unsupervised learning approach removes the need to annotated corpus. Learns the relations from the training data. • Requires considerable large unlabeled domain corpus to generate word embeddings.	(Kang et al. 2014; Wang et al. 2018; Lee et al. 2020)

8.4 PROCESSING CLINICAL NARRATIVE

Clinical narratives are authored by HCPs as part of clinical reports, diagnostic reports, case reports, and patient health records. Clinical narratives contain summary of important health information about patients; history, treatment given, health condition, risks, etc. Following is an example of input and output requirement of clinical narrative processing.

8.4.1 CASE REPORT EXAMPLE

Gurulingappa et al. (2012) A 44-year-old man taking naproxen for chronic low back pain and a 20-year-old woman on oxaprozin for rheumatoid arthritis presented with tense bullae and cutaneous fragility on the face and the back of the hands (Table 8.5).

Concept Extraction & Relation Extraction Task goal:
- **ADE:** cutaneous fragility, tense bullae
- **Drug:** oxaprozin, naproxen
- **Indication:** rheumatoid arthritis, chronic low back pain
- **Patient Info:** 20-year-old woman, 44-year-old man
- **Drug-ADE Pairs:** oxaprozin-cutaneous fragility, naproxen-cutaneous fragility, oxaprozin-tense bullae, naproxen-tense bullae

8.5 PROCESSING PATIENT NARRATIVE

Patients are increasingly using technology and social medial platforms to ask or share information about their medical conditions and treatments. Patient-authored social media posts have the potential to detect ADR in early phases. Patients leverage Health Communities like DailyStrength, PatientLikeMe, MedHelp, and Social media platforms like Twitter and Facebook community group specific to a drug or disease, created by drug manufacturers.

Patients authored input have no or minimal use of standard biomedical concepts or mentions of medical terms describing their medical conditions or symptoms. Input also will contain significant noise caused by repeated characters ("lottttttt of pain"), abbreviations, spelling mistakes, use of special characters, etc. Considerable pre-processing of input text is necessary to remove noise from the patient authored input in comparison to the HCP authored input. Patient use casual and informal language to describe the medical conditions, there is huge variation in describing the similar medical condition by different group of patients. And the process of explaining medical condition in informal words is highly influenced by the patient attributes like demographics, culture, location, occupation, etc. (Table 8.6).

To extract ADR mentions and map it to biomedical standard Lexicons, process starts with building custom dictionary, mapping colloquial input to biomedical standard terms. (Harpaz et al. 2012; Yates and Goharian 2013) Various ADR lexicon-based approach proposed by researchers which maps the colloquial text to standard ADR lexicons from UMLS and MedDRA. Although preparing ADR

TABLE 8.5
Clinical narrative processing approaches

Approach	Input /Annotations	Key Points	Publication Reference
Extracting relations between drugs and adverse effects in MEDLINE case reports.	Manually annotated corpora using ontology driven methodology (OntoEIM).	Java based Relation Extraction (JSRE) uses SVM for classification.	(Gurulingappa et al. 2012)
Annotating and Extracting information from MEDLINE Abstracts.	Manually created ontology: first extracting drug names and then getting additional information from UMLS for medicines, ingredient, and pharmacologic classes.	Proposed approach called AnnoPharma. AnnoPharma is based on three main steps: selection and cleaning step, dictionaries and ontology building step and annotation step. Uses morphological analysis of the extracted words.	(Benzarti and Abdessalem Karaa 2013)
Deep learning-based sequence labeling approach to detect attributes of medical concepts in clinical text.	Sequence labeling approach using Bi-LSTM-CRFs on the attribute detection task.	Bi-LSTMs-CRF architecture to detect various medical concept-attribute pairs.	(Xu et al. 2019)
Co-occurrence based approach to determine relation between entities and acquiring association between biological and clinical entities.	Manually created gold annotations.	Bi-LSTM-CNN and CRF for final classification. Chi-square to evaluate the significance of the disease-drug association.	(Kang et al. 2014; Wang et al. 2018)
Drug-ADE Relation Extraction, Drug-Reason relation extraction.		Weighted BiLSTM, Inter Sentence, Intra Sentence. Walk-based algorithm.	(Kang et al. 2014)

mapping lexicon that incorporate all possible informal phrases to describe a medical condition is impracticable. (Ginn et al. 2014; Sarker and Gonzalez 2015; Akhtyamova et al. 2017; Rezei et al. 2020) Researchers proposed ways to classify probable segments of text containing ADR mentions which potentially reduce the vastness of ADR mapping Lexicon. (Ma and Hovy 2016; Tiftikci et al. 2017) proposed ways to use neural network architecture trained on Biomedical corpora for detecting ADR mentions in the text, ADRs will be extracted from the mentions

TABLE 8.6

Complexities in processing patient narrative

Complication	Examples
Ambiguity	• Example-1: Stay awake is a medicine, but this is a general phrase used by patients to denote a state.
	• Example-2: Low fat yogurt contains tons of protein and calcium, which can fight hunger, stimulate weight loss by burning fat. This mentions calcium and weight loss but it is not indicating an ADR.
	• Example-3: If allergy is mentioned with a drug name, was the user taking the drug to treat existing allergy symptoms, or did allergy symptoms appear as a side effect of taking the drug?
Use of Creative expressions	• Use of acronyms like (I'm, I've).
	• Use of repeated characters: pleaseeeeeeeeeee and/or yessss.
	• Frequent use of idiomatic, ambiguous, and sarcastic expressions.
	• Use of ambiguous/non-standard terms for expressing adverse reactions (e.g., "look like a zombie", "ton of weight").
	• "Allergies: [Bactrim]Drug ([rash]ADE)," the relation between ADE and Drug is not evident as there are no keywords to support.
Noise	• Grammatical and spelling errors.
	• Use of special characters such as @,!, /, *, $, etc.
	• Use of URLs, emoji, and hashtags.

using rule-based approach subsequently. Considering the unimaginable number of variations possible for mentioning the ADR, accuracies of detecting the ADR from informal text is challenging. Alabi et al. (2020) Researchers proposed way to validate these extracted ADRs using Biomedical Knowledge graphs. Table 8.7 consolidates the various approaches suggested by researchers to perform clinical NLP on patient narrative.

8.6   PROCESSING LOW RESOURCE LANGUAGE NARRATIVES

As we have seen so far, common approach towards biomedical concept identification, extraction, normalization, and ADR detection is to either have dictionaries/ standards like UMLS and MedDRA available for concept matching or having large amount of training data to learn representations for biomedical concepts.

Low resource languages do not have amount of data required to train the Deep learning models nor have the standards/ontologies available to normalize the extracted concepts. As we have learned that Word embeddings generated using non-domain text corpus do not perform well on the domain tasks therefore, using word embeddings models like Word2Vec, BERT trained on common non-domain corpora of low resource languages will not be helpful in detecting ADRs. Non-availability of domain corpora in low resource languages is a major challenge to build ADR models for these low resource languages.

TABLE 8.7

Patient narrative processing approaches

Approach	Input/Annotations	Details	Publication Reference
Manual annotations with Semi automated techniques for ADE Classification of the text	Manually created dictionaries for mapping colloquial text to ADE mentions	Tree-based Dictionary matching algorithm to classify the ADR mentions and then finding the standard mapping to MedDRA.	(Harpaz et al. 2012)
Pattern based ADR Extractions	Manually created annotations and dictionary for concept matching.	Created handcrafted patterns for ADR Extraction. Extracted ADR to be matched in the dictionary for normalization. Dictionary is created manually and enhanced using UMLS.	(Yates and Goharian, 2013)
Word embedding based social media text classification for ADR	ADR extraction from Twitter, "Ask a patient" website comments	Applying pre-trained Word2vec or FastText word embeddings for CNN based user comment classifications	(Akhtyamova et al. 2017; Rezei et al. 2020)
Word embeddings & ML classification based combined approach to ADR Classification	ADR extraction from DailyStrength community, Twitter ADE Corpus(EHR s reports), Gold Annotated Twitter data of 76 Drugs (Benzarti and Abdessalem Karaa, 2013)	• Text feature generation sentiment, polarity, Topics, concepts. • ADR assertive text segments can be automatically classified from text-based data sources. • ADR and Non-ADR classification using SVM, NB, Maximum Entropy	(Ginn et al. 2014; Sarker and Gonzalez, 2015)
Rules/Dictionary based approach combined with Deep learning	FEARS Drug labels listing the known ADRs of the Drug	• CNN+Bi-LSTM+ CRF Architecture for classifying and extracting ADR mentions. • Uses Character embeddings (learned using CNN) and Word Embeddings	(Ma and Hovy, 2016; Tiftikci et al. 2017)

(Continued)

TABLE 8.7 (Continued)
Patient narrative processing approaches

Approach	Input/Annotations	Details	Publication Reference
		(Word2Vec) trained on Biomedical corpora. • Soysal et al. (2018) Applying Rule based approach for entity normalization.	
Knowledge Graph based concept relation validations for ADR extraction	Uses a dictionary-based concept recognition and normalization tool, developed at the Erasmus University Medical Center	• Uses existing Dictionary based system for concept identification. Apply Rule based NLP for coordination resolution, abbreviation expansion, term variation, boundary correction, and concept filtering. • Identified concepts are validated using knowledge graph (KG). KG is created using UMLS	(Alabi et al. 2020)

Normalizing the extracted mentions is commonly performed using dictionary-based matching with standard thesaurus/ontologies like UMLs & MedDRA. These standards are currently available only in seven languages. More than 310 languages exist with at least 1 million L1-speakers each (Arizona State University). Wikipedia exists for 300 languages, an indicator for the use of digital infrastructure in these languages (Hedderich et al. 2020). Africa and India are the hosts of around 2,000 low resource languages, and are home to more than 2.5 billion in habitants (Magueresse et al. 2020).

Performing clinical NLP in low resource language settings further complicates the clinical text processing challenges described in earlier sections.

- Unavailability of huge amount of digital data for training deep learning models for domain and non-domain tasks.
- Unavailability of standards, ontologies, thesaurus for biomedical concepts & biomedical concept relations.
- Unavailability of biomedical mappings, Lexicon dictionaries, and labeled data.
- Unavailability of shared clinical dataset like MIMIC or THUME corpus in English.
- Unavailability of Synset and lexical database in the low resource language.

- Unavailability of labeled data for training robust NER tagger, POS tagger, etc.
- Handcrafted rules & patterns for extracting biomedical concepts, entities & relations written for rich resource languages cannot be leveraged as-is for low resource languages.
- Low resource Indian languages like Hindi and Marathi are morphologically rich and agglutinative making it difficult to accurately perform NLP tasks.
- Low resource language ambiguities, ex: ambiguity between common noun and proper noun, ambiguity of a proper name, inflected names, etc.
- Same linguistic rules cannot be applied to all the languages:
  - Work boundary using whitespace demarcation is not standard. Chinese language does not have spaces.
  - There is no provision of capital letters in most of non-English languages to denote nouns.
  - Variations in spellings. ex: Color/Colour, Rama/Ram.
- Unable to leverage established NLP tools/techniques trained on Rich resource language data to perform NLP tasks on low resource languages like POS-taggers, NER taggers, Tokenizers, etc.

In summary all the challenges can be grouped in form of Labeled/Non-Labeled/Standard Data availability, unable to leverage existing NLP tools from rich resource language and complexity of low resource language itself. Tables 8.8 and 8.9 captures commonly used techniques to handle data unavailability challenges and language complexity challenges.

8.7 CLINICAL NLP SYSTEMS

Clinical NLP systems are developed to extract various type of information related to patient, patient medical history, treatment information, and temporal information from the clinical text. These systems are able to successfully detect and categorize patient information related to a specific disease and symptoms. However, these systems require extensive domain specific training and customization efforts to produce desired results, also their performance degrade significantly when introduced to unknown domain settings.

Clinical NLP pipeline requires some specific clinical language processing modules in addition to common NLP pipeline to extract clinically relevant information from clinical text. Table 8.10 details NLP modules required to perform clinical NLP.

Various clinical NLP systems have developed and deployed targeting specific domain and are proven to be successful. These systems are trained using enormous biomedical corpora from PubMed articles, Wikipedia, etc. They have built large expert annotated gold standard datasets, and some have developed expert review and expert feedback framework to iteratively enhance, learn, and improve based on human feedback. Table 8.11 captures details on some of widely used application/NLP systems in the clinical NLP space.

TABLE 8.8

Approach towards solving low resource language data unavailability

Approach	Details	Key Points	Publication Reference
Unlabeled, Labeled and Gold data Creation	• Manual creation of dictionaries or Gold dataset with the help of human domain experts. • Various crowd sourcing methods used for labeling or tagging data. • Semi-automated techniques to generate more data from the existing small corpus by leveraging grammar patterns, synonyms, entities from the same group ex: replacing city names etc. • Leveraging parallel corpora to generate labeled data for low resource language, by mapping the labels from rich resource language to low resource language tokens.	• Time-consuming and costly. • Difficult to resolve conflicts and accuracy issues in case of crowd sourcing.	(Magueresse et al. 2020)
Statistical Machine Translations	• Dictionary having parallel mapping of words from rich resource language to low resource language.	• Timing-consuming and expensive to create parallel dictionary for each source-target language pairs. • Perform poorly on unseen words.	
Neural Machine Translation	• Commercial and free translation services exist for generating machine translations from one language to another for ex: Google Translate, Microsoft Bing Translation services. Facilitating translations from multiple source and target languages.	• Google Translate accuracy varies drastically from one source-target pair to another. Specifically, for low resource language pairs. • Transfer learning works best when parent (HRL) and child (LRL) have similar lexicon and grammar.	(Sennrich et al. 2015; Koehn and Knowles 2017; Kumar et al. 2019; Alabi et al. 2020; Baziotis et al. 2020; Koehn 2017, Aagaard et al. 2012)

- Leveraging data available in the closely related high resource language for training models that can be leveraged by low resource language.
- Transfer learnings from Rich resource to related high resource language (HRL) translations, to perform rich resource to low resource language (LRL) translations.
- Zero-shot learning trains the model using multiple languages and dialects simultaneously. Model will use the learned language pairs to translate between unseen pairs. For ex: Learning from English-French and French-Chinese to perform on unseen pair English-Chinese.
- [83] NMT model capable of open-vocabulary translation by encoding rare and unknown words as sequences of subworld units.

- (Sennrich et al. 2015) NMT system only outperforms their Phrase Based statistical MT system when more than 100 million words (approx. 5 million sentences) of parallel training data are available.
- Unsupervised machine translational do not work when the languages are morphologically different or when training domains are totally different.
- Deep Multi-layer architecture used for rich resource languages having large amount of training data does not perform well on low resource language settings. Shallow layers proven to perform better for low resource language with less data.

Biomedical Reference Corpora Generation

- Integration of terminologies and ontologies in UMLS or SNOMED-CT (Standard Thesaurus) for low resource languages.
- Generated Corpora is French for de-identification, drug-drug interaction in Spanish.

(Sennrich et al. 2015; Dalianis and Velupillai, 2010; Grouin et al. 2014)

(Continued)

TABLE 8.8 (Continued)
Approach towards solving low resource language data unavailability

Clinical Tool language Adaptation	• Annotated corpora generated for training statistical methods. • Clinical NLP tool (cTAKES) working well for Clinical NLP in English language adapting to languages like Spanish. • Combining machine translation and language-specific UMLS resources to use cTAKES for clinical concept extraction from German.		(Grouin et al. 2014; Becker and Böckmann 2016)
Biomedical Domain Adaptation	• Pre-trained word embedding generated for using large non-domain data adapted to biomedical task. And then compared to embedding generated using small-domain curated corpora.	• Pre-trained low resource language embeddings generate using general corpora is not useful for performing Biomedical domain NLP tasks.	(Alabi et al. 2020)
International Community Efforts for generating Multilingual clinical datasets	• The NTCIR-11 MedNLP-2 and NTCIR-12 MedNLPDoc tasks focused on information extraction from Japanese clinical narratives to extract disease names and assign ICD10 codes to a given medical record. • The CLEF-ER 2013 evaluation lab was the first multi-lingual forum to offer a shared task across languages	• Aramaki et al. (2016) Resulted in a small multi-lingual manually validated reference dataset. • Rebholz-Schuhmann et al. (2013) prompted the development of a large gold-standard annotated corpus of clinical entities for French	(Chapman et al. 2011; Rebholz-Schuhmann et al. 2013; Aramaki et al. 2014; Névéol et al. 2014; Kors et al. 2015; Aramaki et al. 2016)

TABLE 8.9
Approach towards solving low resource language complexities

Approach	Details	Pros/Cons	Publication Reference
Learning Language model based on character-level inputs	• Language modeling with character-level input, using CNN+LSTM+RNN architecture.	• Performs better and have significantly less parameters as the input are not word embeddings. • Suitable for morphological rich low resource languages.	(Kim et al. 2016)
Character Sequence Embedding to capture language specific characteristics like morphology	• CHARAGRAM embedding addresses differences due to spelling variations, morphology, and word choice. • Embeds character sequence by adding the vectors of its character n-grams. • Character ngrams reveals etymological links; e.g., die is close to mort	CHARAGRAM model is able to learn precise vectors for particular character n-grams with specific meanings.	(Wieting et al. 2016)
Subword Segmentation to generate embedding for unseen words	• Byte Pair Encoding (Sennrich et al. 2015): Learning subword vectors based on the frequency of the subword in the corpus. For ex: for the word "Lowest" two embeddings will be generated "LOW" and "EST". • WordPiece (Schuster and Nakajima 2012): Generating subword vectors based on the likelihood. Technique is used for generating BERT word embeddings. • SentencePiece (Kudo and Richardson 2018): a language-independent subword tokenizer and detokenizer learned from raw sentences. • Unigram Language Model: subword regularization, which trains the model with multiple subword segmentations probabilistically sampled during training.	• Able to handle open-vocabulary and unseen words. • Kudo and Richardson (2018) will be able to learn embeddings for non-segmented languages such as Korean, Japanese and Chinese. • Kudo (2018) report improvement in low resource language and out-of-domain settings.	(Schuster and Nakajima 2012; Sennrich et al. 2015; Kudo and Richardson, 2018; Kudo, 2018; Song et al. 2019)

TABLE 8.10
Clinical NLP modules

NLP Module	Approach	Challenges
Sentence Boundary detection	• Goal is to detect sentence boundaries in the given text and extract individual sentences. • Machine learning based sentence classifiers like SVM are trained. • Pre-defined Rule/Patterns are configured by experts.	• In western languages "." and "," delimits the sentence boundary but these symbols are also used as part of abbreviations "etc.", dates "12.04.2021", providing medical information "0.5 mg" in the clinical text. • In most of the languages we have some symbols delimiting the sentences but languages like Chinese are exceptions and will need different approach towards sentence detection.
Tokenization	• Detect word tokens from extracted sentences that can be processed and labeled in next steps.	• Díaz and Maña López (2015) collated language complexities like hyphenated compound words (x-ray), words combined with periods, numbers, brackets, punctuations (10%) increase the complexity of extracting word token from the sentences. • Díaz and Maña López (2015) explained biomedical language complexities like DNA sequencing (CCGAAACTGAAAGG), temporal expressions, drug names (x-trozine), protein names (CaMKIV/Gr).
Acronym & Abbreviation processing & Disambiguation	• Identifying and normalizing the acronym and abbreviations found in the text. • Capital letters (Dr.), adjectival expressions (CD8+ for CD8-positive), CT Scan, CHF, etc. • Lexical database is created or standard references or metathesaurus like UMLS used to identify the acronyms/abbreviation.	• Having an updated lexicon covering entire acronyms is impractical. • Resolving conflicts using the current context in case of similar acronyms.
Part-of-speech-tagger	• Assigning relevant PoS tags to the tokens.	Requires annotated domain corpus to deliver expected results.

	• Various machine learning models are trained on biomedical corpus for pos tagging.	
Named Entity recognizer	• Extracting named entities like person, org, dates, phone, names, numbers, etc. • Machine learning models using CRF, dictionary-based approach and Regular expression-based approaches are widely used for NER.	Requires annotated domain corpus to deliver expected results.
Assertion & Negation	• Extracting presence, absence, conditionally present, hypothetically present or mention of medical condition, symptom, disease, etc. in the given clinical text. • Extracting assertion and negation is two step process. First, we have to detect the mention of medical concept/entity and secondly extract the assertion/negation on the concept/entity. Ex: patient **denies pain** in the **heart**. • Various regular expression-based, machine learning and deep learning-based methods are used to extract assertion and negations from the text.	Patients use creative expression to describe the medical conditions, which complicated the assertion/negation detection. Example: I slept like a baby. It was killing me all night.
Medical Coding	• Mapping the mentions of medical procedures, diagnosis in the text to standard medical terminology. • Medical code sets are made available by organizations like WHO, American Medical Association(AMA), Centers for Medicare and Medicaid services (CMS), MedDRA etc.	• Medical code database updates frequently. • Medical mentions in the text maps to multiple medical codes. • Non-rule-based approach towards resolving conflicts while maintaining the prediction accuracy is difficult.

TABLE 8.11

Clinical NLP systems & applications

NLP System/Application	Details
MedLEE (Friedman et al. 1994)	• A Medical Language Extraction and Encoding System (MedLEE). • MedLEE extracts structures and perform medical encoding. • Tool include NLP components to preprocess lexicon, abbreviations, document section names, and context rules, categorizer to categorize words into semantic categories, composer to create relevant phrase from words, and an encoder for medical encoding.
MetaMap/MetaMap Lite (Aronson and Lang, 2010)	• Tool to retrieve biomedical text from MEDLINE/PubMed citations. • Enabling linking between biomedical literature and knowledge. • Provides features to generate variants of the words and phrases in the text and find the results that best match text in the Metathesaurus. • Uses lexical filtering, manual filtering and syntactic filtering to remove noise from the text.
cTAKES (Savova et al. 2010)	• clinical Text Analysis and Knowledge Extraction System builds on Unstructured Information Management Architecture (UIMA) and Open NLP. • cTAKES NLP components are trained using gold standard datasets created from clinical notes from the Mayo Clinic EMR. • cTAKES currentl have NLP components: Sentence boundary detector, Tokenizer, Normalizer, Part-of-speech (POS) tagger, Shallow parser, Named entity recognition (NER) annotator, including status and negation annotators. • Uses SPECIALIST Lexical Tools used for normalization of each word in the input text. • Uses dictionary created from UMLS, SNOMED-CT and RxNORM for concept terminology.
MedTagger (Liu et al. 2013)	• MedTagger is collaborative framework that helps in aggregation and labeling of medical scans dataset using crowdsourcing. • Provides mechanism for label validation.
CLAMP (Soysal et al. 2018)	• CLAMP Java application build on UIMA framework. • CLAMP provides NLP component based on machine learning and rule-based approaches. • CLAMP provides customizable component to perform NLP tasks like sentence boundary detection, tokenization, section heading taction, Pos tagging, NER, medical coding etc. • CLAMP provide GUI based interface for corpus annotation and training.

8.8 SUMMARY

Drug safety is of critical importance to improve healthcare. ADR reporting is one of the critical aspects of drug administration and regularization. Considering various challenges in processing ADR mentions from clinical text to assist ADR reporting procedure, advanced techniques need to be introduced that can significantly reduce the data needs and expert efforts. Domain knowledge is to be made available, international community efforts need to develop multi-lingual clinical datasets, and reference standards need to encourage ADR reporting for drug safety in developing nations.

REFERENCES

Aagaard, L., J. Strandell, L. Melskens, P. Petersen, and E. Hansen. 2012. Global patterns of adverse drug reactions over a decade: Analyses of spontaneous reports to VigiBase™. *Drug Safety: An International Journal of Medical Toxicology and Drug Experience.* 35. 10.2165/11631940-000000000-00000

Airola, A., S. Pyysalo, J. Björne, T. Pahikkala, F. Ginter, and T. Salakoski. 2008. All-paths graph kernel for protein-protein interaction extraction with evaluation of cross-corpus learning. BMC Bioinformatics 2, no. S2 (2008). https://doi.org/10.1186/1471-2105-9-S11-S2

Akhtyamova, L., M. Alexandrov, and J. Cardiff. 2017. Adverse drug extraction in twitter data using convolutional neural network. 2017 28th International Workshop on Database and Expert Systems Applications (DEXA), Lyon, pp. 88–92, doi: 10.1109/DEXA.2017.34

Alabi, J., et al. 2020. Massive vs. Curated embeddings for low-resourced languages: The case of Yorùbá and Twi. Proceedings of The 12th Language Resources and Evaluation Conference. European Language Resources Association: Marseille, France.

Aramaki, E., M. Morita, Y. Kano, and T. Ohkuma. 2014. Overview of the NTCIR-11 MedNLP-2 task. In Proceedings of the 11th NTCIR Conference. Tokyo Japan.

Aramaki, E., M. Morita, Y. Kano, and T. Ohkuma. 2016. Overview of the NTCIR-12 MedNLPDoc task. In Proceedings of the 12th NTCIR Conference on Evaluation of Information Access Technologies. Tokyo Japan.

Arizona State University. "Diego Lab", http://diego.asu.edu/index.php?downloads=yes

Aronson, A. R. and F.-M. Lang. 2010. An overview of metamap: Historical perspective and recent advances. *Journal of the American Medical Informatics Association* 17, no. 3: 229–236.

Baziotis, C., B. Haddow, and A. Birch. 2020 "Language Model Prior for Low-Resource Neural Machine Translation." arXiv preprint arXiv:2004.14928

Becker, M., and B. Böckmann. 2016. Extraction of umls®concepts using apache ctakesTM for german language. Stud Health Technol Inform 223: 71–76. PMID: 27139387.

Benzarti, S., and W. B. Abdessalem Karaa. 2013. AnnoPharma: Detection of substances responsible of ADR by annotating and extracting information from MEDLINE abstracts. 2013 International Conference on Control, Decision and Information Technologies (CoDIT). IEEE.

Buyko, E., E. Beisswanger, and U. Hahn. 2012. *The Extraction of Pharmacogenetic and Pharmacogenomic Relations–A Case Study Using PharmGKB.* Pac Symp Biocomput. Hawaii, USA. Singapore: World Scientific, 376–387.

Chapman, W. W., P. M. Nadkarni, L. Hirschman, L. W. D'Avolio, G. K. Savova, and O. Uzuner. 2011. Overcoming barriers to NLP for clinical text: The role of shared tasks and the need for additional creative solutions. *Journal of the American Medical Informatics Association* 18, no. 5: 540–543.

Crichton, G., S. Pyysalo, B. Chiu, and A. Korhone. 2017. A neural network multi-task learning approach to biomedical named entity recognition. *BMC Bioinformatics* 18, no. 1: 368.

Dalianis, H., and S. Velupillai. 2010. De-identifying Swedish clinical text-refinement of a gold standard and experiments with Conditional random fields. *Journal of Biomedical Semantics* 1, no. 1: 6.

Demner-Fushman, D., W. Chapman, and C. McDonald. 2009. What can natural language processing do for clinical decision support? *Journal of biomedical informatics* 42, no. 5: 760–772. https://doi.org/10.1016/j.jbi.2009.08.007

Denny, J. C., P. R. Irani, F. H. Wehbe, J. D. Smithers, A. Spickard III. 2003. The KnowledgeMap project: development of a concept-based medical school curriculum database, *AMIA Annu Symp Proc.* 2003: 195–199.

Díaz, N. P. C., and M. J. Maña López. 2015. An analysis of biomedical tokenization: Problems and strategies. In Proceedings of the Sixth International Workshop on Health Text Mining and Information Analysis, pp. 40–49, Lisbon, Portugal. Association for Computational Linguistics.

European Medicine Agency. 1995. "EudraVigilance", https://www.ema.europa.eu/en/human-regulatory/research-development/pharmacovigilance/eudravigilance

Friedman, C., P. O. Alderson, J. H. M. Austin, J. J. Cimino, and B. Stephen Johnson. 1994. A general natural-language text processor for clinical radiology. *Journal of the American Medical Informatics Association* 1, no. 2: 161–174.

Fundel, K., R. Küffner, and R. Zimmer. 2007. RelEx—Relation extraction using dependency parse trees. *Bioinformatics* 23, no. 3: 365–371.

Ginn, R., P. Pimpalkhute, A. Nikfarjam, A. Patki, K. O'Connor, A. Sarker, et al. 2014. Mining twitter for adverse drug reaction mentions: A corpus and classification benchmark. Fourth workshop on Building and Evaluating Resources for Health and Biomedical Text Processing (BioTxtM). Reykjavik, Iceland.

Government of Canada. 2020. "Drugs and health products", https://www.canada.ca/en/health-canada/services/drugs-health-products.html

Grouin, C., T. Lavergne, and A. Névéol. 2014. Optimizing annotation efforts to build reliable annotated corpora for training statistical models. In Proceedings of LAW VIII - The 8th Linguistic Annotation Workshop, pp. 54–58. Dublin, Ireland. Association for Computational Linguistics and Dublin City University.

Gurulingappa, H., et al. 2012. Development of a benchmark corpus to support the automatic extraction of drug-related adverse effects from medical case reports. *Journal of Biomedical Informatics* 45, no. 5: 885–892.

Gurulingappa, H., A. Mateen-Rajpu, and L. Toldo. 2012. Extraction of potential adverse drug events from medical case reports. *Journal of Biomedical Semantics* 3, no. 1: 1–10.

Habibi, M., L. Weber, M. Neves, D. L. Wiegandt, and U. Leser. 15 July 2017. Deep learning with word embeddings improves biomedical named entity recognition. *Bioinformatics* 33, no. 14: i37–i48, doi: 10.1093/bioinformatics/btx228

Harpaz, R., W. DuMouchel, N. H. Shah, D. Madigan, P. Ryan, and C. Friedman. 2012. Novel data-mining methodologies for adverse drug event discovery and analysis. *Clin Pharmacol Ther.* 91, no. 3: 1010–1021. [PubMed: 22549283].

Hedderich, M. A., et al. 2020. "A Survey on Recent Approaches for Natural Language Processing in Low-Resource Scenarios." arXiv preprint arXiv:2010.12309

Huang, Y., H. J. Lowe, D. Klein, and R. J. Cucina. 2005. Improved identification of noun phrases in clinical radiology reports using a high-performance statistical natural language parser augmented with the UMLS specialist lexicon. *J Am Med Inform Assoc.*

2005 May-Jun; 12, no. 3: 275–285. doi: 10.1197/jamia.M1695. Epub 2005 Jan 31. PMID: 15684131; PMCID: PMC1090458.

Kandula, S., and Q. Zeng-Treitler. 2010. Exploring relations among semantic groups: A comparison of concept co-occurrence in biomedical sources. *Stud Health Technol Inform* 160: 995–999.

Kang, N., et al. 2014. Knowledge-based extraction of adverse drug events from biomedical text. *BMC Bioinformatics* 15, no. 1: 1–8.

Katrenko, S., and P. Adriaans. 2006. Learning relations from biomedical corpora using dependency tree levels. KDECB'06 Proceedings of the 1st International Conference on Knowledge Discovery and Emergent Complexity in Bioinformatics; Ghent, Belgium. Heidelberg: Springer, pp. 61–80.

Kim, Y., Y. Jernite, D. Sontag, and A. Rush. 2016. Character-aware neural language models. *Proceedings of the AAAI Conference on Artificial Intelligence*, 30, no. 1. Retrieved from https://ojs.aaai.org/index.php/AAAI/article/view/10362

Koehn P., and R. Knowles. 2017. Six challenges for neural machine translation. In Proceedings of the First Workshop on Neural Machine Translation, pp. 28–39, Vancouver.

Kors, J., S. Clematide, S. Akhondi, E. van Mulligen, and D. Rebholz-Schuhmann. 2015. A multilingual gold-standard corpus for biomedical concept recognition: The Mantra GSC. *J Am Med Inform Assoc* 22, no. 5: 948–956.

Kudo, T. 2018. Subword regularization: Improving neural network translation models with multiple subword candidates. In Proceedings of the 56th Annual Meeting of the Association for Computational Linguistics. (Volume 1: Long Papers), pp. 66–75, Melbourne, Australia. Association for Computational Linguistics.

Kudo, T., and J. Richardson. 2018. "Sentencepiece: A simple and language independent subword tokenizer and detokenizer for neural text processing." arXiv preprint arXiv:1808.06226

Kumar, R., P. Jha, and V. Sahula. 2019. An Augmented Translation Technique for Low Resource Language Pair: Sanskrit to Hindi Translation. Proceedings of the 2019 2nd International Conference on Algorithms, Computing and Artificial Intelligence, December 2019, pp. 377–383.

Lafferty, J., A. McCallum, and F. Pereira. 2001. Conditional random fields: Probabilistic models for segmenting and labeling sequence data. Proceedings of the 18thInternational Conference on Machine Learning 2001 (ICML 2001), pp. 282–289.

Lee, J., et al. 2020. BioBERT: A pre-trained biomedical language representation model for biomedical text mining. Bioinformatics 36, no. 4: 1234–1240.

Liu, H., S. J. Bielinski, S. Sohn, S. Murphy, K. B. Wagholikar, S. R. Jonnalagadda, K. Ravikumar, S. T. Wu, I. J. Kullo, and C. G. Chute. 2013. An information extraction framework for cohort identification using electronic health records. *AMIA Summits on Translational Science Proceedings* 2013: 149.

LVG. 1994. https://lhncbc.nlm.nih.gov/LSG/Projects/lexicon/current/web/index.html

Ma, X., and E. Hovy. 2016. "End-to-end sequence labeling via bi-directional lstm-cnns-crf." arXiv preprint arXiv:1603.01354

MacKinlay, A., H. Aamer, and A. J. Yepes. 2017. Detection of adverse drug reactions using medical named entities on Twitter. AMIA Annual Symposium Proceedings. Vol. 2017. American Medical Informatics Association.

Magueresse, A., V. Carles, and E. Heetderks. 2020. "Low-resource Languages: A Review of Past Work and Future Challenges." arXiv preprint arXiv:2006.07264

McCallum A., D. Freitag, and F. Pereira. 2000. Maximum entropy Markov models for information extraction and segmentation. ICML 17, no. 2000: 591–598.

Névéol, A., C. Grouin, J. Leixa, S. Rosset, and P. Zweigenbaum. 2014. The QUAERO French medical corpus: A resource for medical entity recognition and normalization. In Proceedings of BioText Mining Workshop, LREC 2014. BioTxtM 2014. Reykjavik, Iceland, pp. 24–30.

Pyysalo, S., A. Airola, J. Heimonen, J. Björne, F. Ginter, and T. Salakoski. 2008. Comparative analysis of five protein-protein interaction corpora. In BMC Bioinformatics 9, no. 3: 1–11. BioMed Central.

Rebholz-Schuhmann, D., S. Clematide, F. Rinaldi, S. Kafkas, E. van Mulligen, C. Bui, J. Hellrich, I. Lewin, D. Milward, M. Poprat, A. Jimeno-Yepes, U. Hahn, and J. Kors. 2013. Entity recognition in parallel multi-lingual biomedical corpora: The CLEF-ER laboratory overview. In *Information Access Evaluation. Multilinguality, Multimodality, and Visualization. Lecture Notes in Computer Science*, editors P. Forner, H. Müller, R. Paredes, P. Rosso, and B. Stein, 353–367, Springer.

Rezei, Z., et al. 2020. Adverse drug reaction detection in social media by deep learning methods. *Cell Journal (Yakhteh)* 22, no. 3: 319.

Rinaldi, F., et al. 2007. Mining of relations between proteins over biomedical scientific literature using a deep-linguistic approach. *Artificial Intelligence in Medicine* 39, no. 2: 127–136.

Sarker, A., and G. Gonzalez. 2015. Portable automatic text classification for adverse drug reaction detection via multi-corpus training. *J Biomed Inform* 53: 196–207.

Savova, G. K., J. J. Masanz, P. V. Ogren, J. Zheng, S. Sohn, K. C. KipperSchuler, and C. G. Chute. 2010. Mayo clinical text analysis and knowledge extraction system (ctakes): Architecture, component evaluation and applications. *Journal of the American Medical Informatics Association* 17, no. 5: 507–513.

Schuster, M., and K. Nakajima. 2012. Japanese and Korean voice search. 2012 IEEE International Conference on Acoustics, Speech and Signal Processing (ICASSP), Kyoto, pp. 5149–5152, doi: 10.1109/ICASSP.2012.6289079

Sennrich, R., B. Haddow, and A. Birch. 2015. "Neural machine translation of rare words with subword units." arXiv preprint arXiv:1508.07909

Sennrich, R., B. Haddow, and A. Birch. 2015. "Neural machine translation of rare words with subword units." arXiv preprint arXiv:1508.07909

Song, K., et al. 2019. "Mass: Masked sequence to sequence pre-training for language generation." arXiv preprint arXiv:1905.02450

Song, M., H. Yu, and W. Han. 2015. Developing a hybrid dictionary-based bio-entity recognition technique. *BMC medical informatics and decision making* 15, no. 1: 1–8.

Soysal, E., et al. 2018. CLAMP–A toolkit for efficiently building customized clinical natural language processing pipelines. *Journal of the American Medical Informatics Association* 25, no. 3: 331–336.

Tavassoli, N., et al. 2009. Reporting rate of adverse drug reactions to the French pharmacovigilance system with three step 2 analgesic drugs: Dextropropoxyphene, tramadol and codeine (in combination with paracetamol). *British Journal of Clinical Pharmacology* 68, no. 3: 422–426.

Tiftikci, M., et al. 2017. Extracting adverse drug reactions using deep learning and dictionary based approaches. TAC.

Tsai, R. T. H., C. L. Sung, H. J. Dai et al. 2006. NERBio: Using selected word conjunctions, term normalization, and global patterns to improve biomedical named entity recognition. *BMC Bioinformatics* 7, no. 5: 1–14. BioMed Centra. doi:10.1186/1471-2105-7-S5-S11

US Food and Drug Administration. 2016. "Introduction to FDA's MedWatch Adverse Event Reporting Program", https://www.fda.gov/media/95928/download

VA Center of Medication and Safety. 2006. "Adverse drug events, adverse drug reactions and medication errors", https://www.pbm.va.gov/PBM/vacenterformedicationsafety/tools/AdverseDrugReaction.pdf

Wang, Y., et al. 2019. Applications of natural language processing in clinical research and practice. Proceedings of the 2019 Conference of the North American Chapter of the Association for Computational Linguistics: Tutorials (pp. 22–25).

Wang, X., Y. Zhang, X. Ren, Y. Zhang, M. Zitnik, J. Shang, C. Langlotz, and J. Han. 2018. Cross-type biomedical named entity recognition with deep multi-task learning. *Bioinformatics* 35, no. 10: 1745–1752. ISSN = 1367–4803. 10.1093/bioinformatics/bty869. 10.1093/bioinformatics/bty869

Wieting, J., et al. 2016. "Charagram: Embedding words and sentences via character n-grams." arXiv preprint arXiv:1607.02789

Xu, J., Z. Li, Q. Wei et al. 2019. Applying a deep learning-based sequence labeling approach to detect attributes of medical concepts in clinical text. *BMC Medical Informatics and Decision Making* 19, no. 236: 1–8. doi: 10.1186/s12911-019-0937-2

Yates, A., and N. Goharian. 2013. ADRTrace: Detecting expected and unexpected adverse drug reactions from user reviews on social media sites. Proceedings of the 35th European conference on Advances in Information Retrieval: 816–819. Springer, Berlin, Heidelberg.

Yoon, W., C. So, and J. Lee et al. 2019. CollaboNet: collaboration of deep neural networks for biomedical named entity recognition. *BMC Bioinformatics* 20, no. 10: 55–65. doi: 10.1186/s12859-019-2813-6

Zhang, L., L. Wong, H. Ying (Helen), and I. Wong. 2014. Pharmacovigilance in China: Current situation, successes and challenges. *Drug Safety: An International Journal of Medical Toxicology and Drug Experience* 37, no. 10: 765–770. doi: 10.1007/s40264-014 0222-3

https://www.prosperity.com/download_file/view/4193/2025

9 Language- and Domain-Independent Approach to Automatic Detection of COVID-19 Fake News

Ángel Hernández-Castañeda
Cátedras CONACyT, National Council for Science and
Technology, Mexico City, Mexico

Autonomous University of the State of Mexico, Mexico
State, Mexico

*René Arnulfo García-Hernández, Yulia Ledeneva,
and Griselda Areli Matías-Mendoza*
Autonomous University of the State of Mexico, Mexico
State, Mexico

CONTENTS

DOI: 10.1201/9781003138013-9

9.1　INTRODUCTION

The COVID-19 pandemic has become a significant challenge not only for medical areas but also for computer science. At the same time that the virus is spreading around the world, fake news spreads wildly. This misinformation makes people unable to distinguish the correct disease prevention and treatment methods. Therefore, bad practices spread massively among the world population.

Fake news[1] is defined as false stories that appear to be news and have misinformation as the main goal. In turn, misinformation[2] is closely related to deception, which is the act of hiding the truth, commonly to obtain an advantage[3].

Recent studies (Section 9.2) have addressed fake news detection by considering all content of the documents; however, actually, some of the claims in documents may be true because deceptive text does not strictly require that all claims be false.

Therefore, this study proposes an approach that attempts to keep only the key ideas in the documents since the theory is that the secondary ideas are not focused on producing deceptive text. Thus, in this proposed approach, key ideas are those sentences that contain relevant information to transmit the main document message. Then, a document can dispense with secondary ideas; however, the meaning of the authorws message remains.

To extract key sentences, two general processes were proposed: first, extract the general structure of each document by organizing its sentences using a clustering algorithm; and then, obtain the most relevant sentence of each generated cluster by finding the most relevant words in the sentences. To identify the relevant words in a sentence, a topic modeling algorithm was used to obtain the latent topics of documents. That is, the algorithm can identify the most relevant topics and words using a statistical mechanism.

This general process allows the proposed method to classify only the key sentences of each text and determine if a text is deceptive or truthful. Nevertheless, a precise selection of key sentences plays an important role, so our proposed summary process was compared with those of other authors (see Section 9.3) and showed competitive performance. Thus, research results show that the cues to deception are mainly found in the key ideas of a text. In addition, this approach is compared to the one that considers all sentences.

Because the COVID-19 pandemic is a global phenomenon, the news is expected to be written in different languages. Thus, the main goal of this study was to create a language-independent approach; therefore, none of the procedures introduced require a priori information to generate vectors. The term weighting, LDA, doc2vec, TF-IDF, and OHE generate representations by processing the content of the documents themselves; in addition, the evolutionary clustering process uses internal validation indices as a fitness function, and, therefore, knowledge about classes is not required.

Therefore, fake news was analyzed in Arabic, which is a little-explored language in deception detection because it brings different challenges, such as diglossia and its relative complexity in morphology (Rosso et al. 2018). In addition, there are few NLP tools to address these peculiarities because most of them are constructed for the English language.

9.2 RELATED WORK

Simply stated, a lie is defined as the process to say or write something that is not true to deceive someone;[4] thus, it is an intentional effort to transmit something that is not true. Polage (2017) claims that lying involves more cognitive effort than telling the truth and that an increase in cognitive demand can leave cues to deception. NLP and pattern recognition methods attempt to automatically find these cues to identify deceptive text.

However, the process of detecting lies is very complex, both for humans and for machines, since it implies a cognitive process. Ordinary people, for example, have shown approximately 50% accuracy in detecting lies, while some deception detection experts reach approximately 89%, but only in specific contexts (Rosso and Cagnina 2017). This limited human ability to detect deception has motivated the development of computational models to automatically detect lies. These models can be generated using different sources, such as videos, speeches, and facial expressions; however, the difficulty increases when texts are the only source of information due to the lack of features related to the liar, that is, the words in texts are the only information available.

In addition, unlike other classification problems such as polarity detection, the deception detection task lacks explicit cues or words that define a deceptive or truthful text. For example, in a sentiment analysis task, finding the word "smile" in a text indicates that its polarity could be positive; however, there is no explicit set of words that give clues to deceptive text (at least not in a general context).

The main purpose of a deceptive text is to change peoplews opinion about someone or something, making them believe something that a writer does not believe. Therefore, detecting deception in the text is a very important task for many applications, such as detecting if a review of a product or service is written to deceive people or if fake news published on social networks is trying to discredit a candidate in a political campaign.

Several works have been dedicated to the problem of detecting deception in the text; and in each of them, different methods have been implemented to generate features. Each method might be successful to some degree depending on the type of information that is extracted from the text. For instance, in various works that report competitive results, the Linguistic Inquiry and Word Count (LIWC) tool has been applied to extract psycholinguistic-based features (Newman et al. 2003; Schelleman-Offermans and Merckelbach 2010; Toma and Hancock 2012); and other common methods used to extract lexical information are n-grams (Fornaciari and Poesio 2014) and character n-grams (Fusilier et al. 2015). In addition, syntactic information has also been shown to provide information relevant to deception detection (Feng, Banerjee and Choi 2012, Xu and Zhao 2012).

Recently, the pandemic has provided a wealth of information about the coronavirus that can be false. Misinformation might be harmful because people can take wrong steps to prevent contagion. Thus, NLP researchers have proposed several models to detect this type of information and prevent it from spreading.

9.2.1 DETECTION OF RUMORS

The work of Kar et al. (2020) proposed a basic scheme for COVID-19 fake news detection. This approach consists of a method for mapping features and the use of different classifiers that learn those features to identify fake tweets. The authors extract textual and statistical information from tweet text messages and analyze their role in the classification process.

The authors designed two datasets for the task of detecting fake tweets. These datasets briefly described below consist of tweets in English and Indic languages.

- The Infodemic Covid-19 dataset is a collection of 504 English tweets and 218 Arabic tweets annotated with fine-grained labels related to disinformation about COVID-19. The labels answer seven different questions, but the authors only consider one of them, which is the following: "Does the tweet contain a verifiable factual claim?"
- Two Indic languages were considered by the authors: Hindi and Bengali. First, to obtain the Bengali tweets, the authors randomly selected and annotated 100 tweets from a database of over 36,000 COVID-related tweets. Then, for Hindi, they use a tweepy API to crawl COVID-related tweets from the Twitter interface. This was achieved by firing an API search with COVID-related key terms such as "Covid", "Corona", etc. Finally, the tweets were annotated as fake or non-fake.

A set of feature generation methods and classifiers were proposed by the authors to detect fake tweets. First, they extract various textual and statistical information from the tweet text messages and user information separately and analyze their role in the classification process. The different features are briefly defined as follows:

- Text features: Some of the Twitter and textual features, such as the number of times a tweet was retweeted, the number of likes, and the numbers of upper characters, questions, and exclamation marks in the tweet, were extracted.
- Twitter user features, such as whether the user account was verified officially, the total number of tweets tweeted by the user, the number of days since the user's latest tweet, etc., were extracted from the user profiles.
- Fact verification score: The authors used the tweet text as a search query in the Google search engine and considered the search results from reliable sources only. The authors obtain the fact verification score by taking the average of the Levenshtein distance between the tweet text and the titles obtained from the search results.
- The bias score is calculated using a pretrained profanity checker to obtain the probability of a tweet containing offensive language and use it as the bias score.
- Source tweet embedding: The BERT embedding of the tweet text is considered a feature.

Different classifiers, such as support vector machines, random classifiers, and multilayer perceptrons, were used for the classification. This approach has achieved an F-score of approximately 89%.

Similar to Kar, in the work of Hamid et al. (2020), two methods were used to generate features: a method based on the word count and a method that obtains semantic information (BERT). A graph neural network was applied to the obtained features to build a model for predicting COVID-19 misinformation tweets.

9.2.2   COVID-19 Fake News Detection

The work of Felber (2021) addressed fake news detection by classifying COVID-19 news in English. Therefore, given a collection of COVID-19-related posts from various social media platforms, such as Twitter, Facebook, Instagram, etc., the method classified the posts as either fake or real.

The dataset analyzed (Patwa et al. 2020) contains 10,700 manually annotated social media posts and articles of real and fake news on COVID-19. The dataset was split into training, validation, and test sets.

The authors conduct an experiment with different preprocessing pipelines and apply classical machine learning algorithms with diverse groups of linguistic features such as n-grams, readability, emotional tones, and punctuation. In addition, a preprocessing step that includes stop word and link removal, lemmatization, stemming, lower case transformation, and XML entity replacement is included in the framework.

9.2.3   Multilingual COVID-19 Deception Detection

In (Shahi and Nandini 2020), authors proposed an approach that attempts to identify COVID-19 fake news. The data were collected from 92 different fact-checking websites and manually annotated. The resultant datasets consist of documents in 40 languages from 105 countries.

As a preprocessing step, the authors applied basic NLP techniques to remove unwanted details from the data such as letter casing, short words, and conduct tokenization. In addition, for data cleaning, a language detector and a spelling corrector were implemented.

The authors split the dataset into two classes: false and other. The false class includes the articles that were labeled false by fact-checker and contains 4132 articles. The other class contains 1050 fact-checked articles.

The experiments were conducted on a subset that contains only English articles, and the authors identified deception with a 0.65 F-measure.

Several works (Bang et al. 2021) have adopted the basic machine learning framework to generate models from a set of labeled instances to identify fake texts on social media.

9.3   THE PROPOSED APPROACH FOR COVID-RELATED FAKE NEWS DETECTION

This study proposes a deception detection approach (see Figure 9.1) that attempts to capture the key ideas in a document to avoid secondary information that could affect the

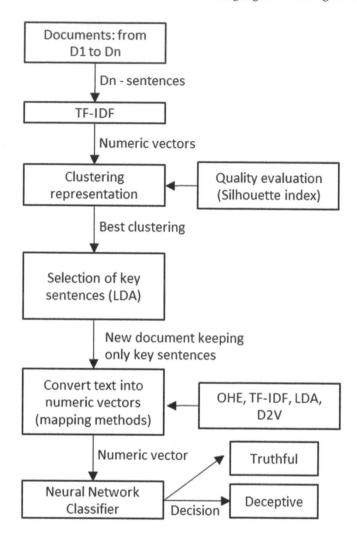

FIGURE 9.1 General framework for fake news detection.

search for cues to deception. To achieve this, a clustering representation is chosen to identify key ideas and form a new collection of synthesized documents that should avoid secondary ideas as much as possible. The clustering representation is built by generating a vector space model using TF-IDF representation, and it is applied to obtain a general structure of the document according to the relevance of its term. Next, the quality of the cluster is evaluated by the silhouette index to obtain the best configuration. Then, relevant sentences are selected from the clusters according to the LDA distribution to form the synthesized document. This process is applied to each document in the collection, and the new documents are converted into a vector representation using the mapping

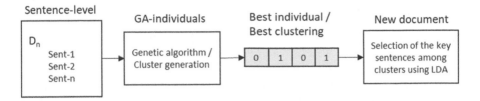

FIGURE 9.2 Summarization process.

methods (Section 9.3.1). Finally, documents are classified as truthful or deceptive. Below is a detailed description of each step of the proposed approach.

As stated before, the main process in this study is to remove the secondary ideas from the original texts of the collections (see Figure 9.2). Then, each document is processed by splitting each document into sentences because the sentence is considered the minimum unit with complete meaning.

Sentences in the split documents are then organized by a partitional clustering representation (detailed in Section 9.3.2). That is, objects are grouped according to their proximity by mapping the problem to a vector space model, where sentences represent the objects in this model. Therefore, each sentence in the source document is mapped to a numerical vector using the TF-IDF scheme. Consequently, this step allows generating groups of sentences according to their relevance in the document and obtaining an overview of the process.

The main problem of partitional clustering algorithms, such as k-means or k-medoids, is to determine the number of groups to be formed since testing each grouping becomes a combinatorial problem. Therefore, this work implements a genetic algorithm (GA) to find the best clustering (detailed in Section 9.3.3) using the silhouette index as a fitness function.

When the GA finds the best clustering configuration, the LDA algorithm is responsible for selecting the key ideas of each cluster (see Section 9.3.4). This is possible because LDA can obtain the word and topic distribution of a source document. This information is used to select the most important sentences as the most likely topics and words are the main components of the document.

As a result, the newly synthesized documents, composed of the sentences selected using the LDA algorithm, contain only the potential key ideas. Finally, these documents are mapped using different feature generation methods, such as OHE, D2V, LDA, and TF-IDF; and they are then classified as truthful or deceptive using a backpropagation neural network.

It is worth noting that the feature generation method does not need any a priori information. For example, first, D2V was trained with the complete current dataset to be analyzed, and then each document was converted into its word embedding representation (inferred vector). Second, LDA, OHE, and TF-IDF do not need any training process to generate vectors. Thus, this proposed approach is language independent because it processes documents without needing any external data but only the terms in the datasets.

9.3.1 MAPPING METHODS

As stated above, the first step of this study requires mapping texts to numeric vectors. Thus, there are various feature extraction methods that obtain information at different language levels.

In this study, various methods were used to measure the importance of features at the lexical and semantic levels. These methods are briefly described below.

9.3.1.1 One-Hot Encoding (OHE)

To build one-hot vectors, a list of all the words W_1, W_2, ..., W_n in the dataset is formed, and then each document is analyzed to determine whether W_n exists in the current text. If so, feature n (F_n) is set to 1; otherwise, it is set to 0.

9.3.1.2 Term Frequency – Inverse Document Frequency (TF-IDF)

TF-IDF attempts to reflect how relevant terms are in a collection of documents by measuring the relevance of a given term with respect to a current document and the entire collection.

9.3.1.3 Latent Dirichlet Allocation (LDA)

LDA (Blei, Ng and Jordan 2003) is an algorithm able to obtain both the latent topics and word distributions of a given document in a collection. In turn, this method also attempts to organize those words in the collection that share a semantic link into groups. Consequently, each document is represented by a distribution of topics, i.e., a vector of probabilities. Each probability represents the degree of belonging of each topic (group of words) to the current document.

Simply put, the LDA generative process consists of the following steps:

1. A determination of the N words a document will contain in accordance with a Poisson distribution.
2. The choice of a mixture of topics for the document according to Dirichlet distribution with respect to a fixed set of K topics.
3. The generation of each word in the document is as follows:
   a. Choosing a topic
   b. Using the chosen topic to generate the word

Using a generative model, LDA analyzes the set of documents to find the most likely set of topics possibly addressed in a document.

We can regard LDA as a source of word features similar to LIWC, but unlike LIWC, LDA generates the groups of words (topics) automatically. Additionally, groups of words from LDA are not labeled, and their contents are different depending on the corpus used to train LDA.

See Figure 9.3 for an example of words that are obtained using LDA.

The Python *Gensim* library was used to obtain the LDA, and hyperparameters were selected as follows. Because a low alpha means that the source document consists of few topics and a low beta means that each topic consists of few words,

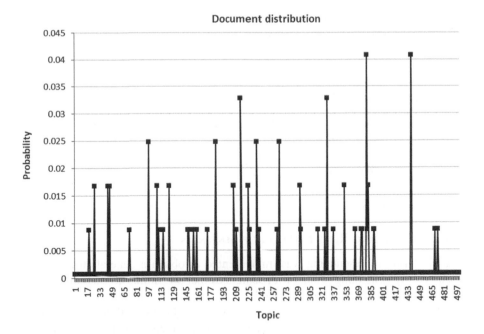

FIGURE 9.3 Topic distribution of a document processed by LDA.

the selection of these hyperparameters was as follows. First, alpha was set to 0.5 because a summary document typically consists of only a few topics. Second, beta was set to 0.001 to obtain a few words on each topic as only the top ten words were useful for key idea selection.

9.3.1.4 Doc2vec

In several studies on machine learning, the authors have searched for numeric representations of the studied objects. Thus, Le and Mikolov (2014) offered a distributed representation of words to build a vector that represents the semantic meaning of each word in a set of documents, considering the context. The goal is to predict a word given the occurrences of other words.

In summary, doc2vec is an unsupervised algorithm that generates fixed-length numeric vectors of variable length text.

doc2vec supplies the word matrix with paragraphs, which provide many sampled contexts (see Figure 9.4). Thus, doc2vec infers new words using the word vector and a vector paragraph, which serve as a memory of the context. It establishes the topic of the document to better predict the next word.

In contrast to the bag-of-words approach, doc2vec can consider the ordering and semantics of the words. In addition, this algorithm avoids sparsity and high dimensionality, in contrast to OHE.

The Python *Gensim* library was used to obtain the doc2vec model using the default hyperparameters. A 100-dimension vector was generated for each document in the collection.

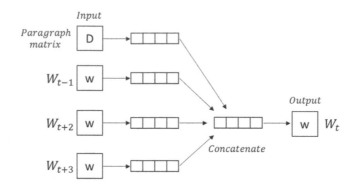

FIGURE 9.4 Doc2vec.

9.3.2 CLUSTERING REPRESENTATION OF SENTENCES

The clustering representation was implemented to obtain a general structure of a document by generating groups with sentences that are similar to each other based on the TF-IDF weighting term method.

The goal of the clustering representation is to avoid the selection of similar ideas to form the synthesized document because a set of sentences with similar meaning might be considered relevant. Thus, selecting relevant sentences through the formed groups attempts to prevent redundancy.

This process involves two aspects: (1) the mapping of sentences to numeric vectors and (2) the selection of proximity measures.

1. The TF-IDF scheme was selected as the weighting term method for mapping each sentence to a numeric vector. This representation with the clustering algorithm is combined to organize similar sentences into groups.
2. A combined Euclidean and cosine measure was used to build the vector space model, and this combination of proximity measures proposed by Gu et al. (2017) was empirically proven to obtain better results in this study. Therefore, to combine both measures and obtain a similarity measure, we applied the following adequacies. First, the Euclidean distance measure was turned into a similarity measure by using the following adequacy: $modifiedEuclidean = \frac{1}{Euclidean + 1}$. Next, the cosine measure was modified by simply adding a unit to obtain only positive values: $modifiedCosine = cosine + 1$ in the range [1,2]. Finally, the proximity between two objects is calculated by $modifiedEuclidean * modifiedCosine$. The resulting measure obtains only positive values that facilitate the representation of the problem using a genetic algorithm representation (Section 9.3.3).

Partitional clustering algorithms assign each object to the closer centroid or attractor; therefore, the clustering representation in this study attempts to assign each sentence (object) to its corresponding group.

However, finding the optimal number of groups to be formed is extremely computationally expensive. Furthermore, using a brute force method to list all possibilities is infeasible as it becomes a combinatorial problem (Xu and Wunsch 2008).

9.3.3 FINDING THE OPTIMAL CLUSTERS

A genetic algorithm (GA) representation is proposed to ensure the covering of sentences in the synthesized document since this document contains only potential key sentences. After this process is applied, each sentence in a document will belong to a cluster based on its word relevance. This process is detailed below.

To solve the problem of finding the optimal number of clusters, an encoding of the problem was necessary, i.e., mapping the real problem (phenotype) to a GA representation (genotype). In this study, an individual or chromosome consists of binary genes wherein each gene represents each sentence in the source document to be synthesized. Thus, the individuals in the population have a fixed length equal to the number of sentences in the source document. A gene value of 1 means that the sentence is a centroid, and 0 means otherwise. Each sentence that is not checked as a centroid should belong to some group formed. Thus, an individual formed of n-centroids would form n-clusters. The sentences marked as the centroids of the clusters are considered the general ideas of the document whereas the sentences attracted by these sentences are considered ideas that are close to the main topic.

The initial population was generated by randomly assigning a value to each gene. That is, given an individual $i = \{g_1, g_2, ..., g_n\}$, each g_n was randomly set to 0 or 1. To generate new individuals, we applied the one-point crossover technique and a standard mutation wherein the operator inverts the binary value of a selected gene.

The allocation of the fitness of individuals over the population was addressed by selecting the silhouette index as a fitness function due to its correlation with key phrase extraction (Hernández-Castañeda et al. 2020). Furthermore, according to (Liu et al. 2010) and (Rendón et al. 2011), the silhouette index performed better than the other indices in clustering problems. The real values considered by this index are between 1 and −1, where the values closest to 1 represent better clustering.

Offspring selection was conducted by using a roulette operator that selects the best individuals with a high probability; however, this operator does not completely discriminate bad solutions, allowing the population to be diverse. Other methods, such as random, ranks, and tournament selection, were also applied in this study; however, they obtained lower performance.

9.3.4 SELECTION OF MAIN SENTENCES USING LDA

The selection of key sentences can be performed by extracting the centroid sentences from the clustering configuration; however, these sentences may not be the most relevant of their group.

For example, centroids should meet the separation property to optimize the cluster configuration. Thus, the objects of a formed group may have relevant terms in common, and these terms should be dissimilar to the other groups. This process

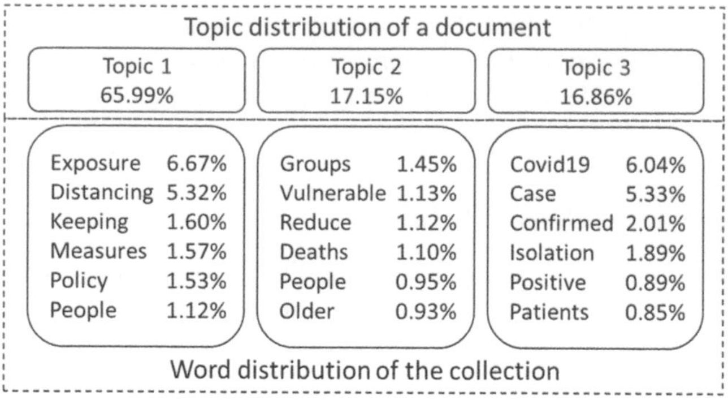

FIGURE 9.5   Document topic distribution.

does not guarantee that the centroid sentences contain the most relevant information but only that the relevant terms are as different as possible. Therefore, key sentences could be some of those attracted by the centroid.

Therefore, in this study, we propose to extract the latent structure of documents using the topic modeling algorithm called Latent Dirichlet Allocation (LDA). Figure 9.5 shows a topic and word distribution obtained from a text. For example, given a sentence, topic 1 has a 65.99% probability of being part of it; and, in turn, the word "exposure" has a probability of 6.67% of being selected into Topic 1.

The LDA model is configured to generate three topics in the experiments of this study because this configuration has been empirically proven to yield the best results. This model is applied in two steps of the proposed approach: in the process of mapping text to numeric vectors by using the topic distribution and in the selection of the key sentences by using the bags of words grouped in each topic (see Figure 9.5).

With the aim of identifying the key sentences, the ten most representative words of three topics are selected as keywords. Thus, given each probability pT_i associated with each topic T_i and each probability P_i associated with each word w_i in the representative sentence, each word w_s in the candidate sentence was compared with w_i. If w_i was equal to w_s, $P_i * pT_i$ was accumulated in *pTotal*. The sentence that reached the maximum value of *pTotal* was selected as a key sentence. If there is no $w_i = w_s$ in any sentence of the cluster, the centroid sentence is selected to be part of the synthesized document. This guarantees that at least one of the sentences of each cluster will be included in the new document. The key sentence selection process ends when a candidate sentence is extracted from each cluster. Therefore, the clustering scheme allows us to cover the topics of the document, and the LDA's word distribution allows us to select the most promising key ideas in each cluster. In addition, if the clusters consist of a single sentence because the source text is very short, only one sentence is selected from all clusters considering the *pTotal* value.

In Table 9.1, an example of the keywords obtained by the LDA model of the original document is shown, that is, those that are statistically relevant in a document; therefore, keywords provide a guide to identifying relevant sentences in each cluster.

TABLE 9.1

Keywords obtained by an LDA model

Topics	Words
Topic 1	Covid-19, coronavirus, cases, deaths, confirmed, total, states, reported, India, samples
Topic 2	people, patients, vaccine, testing, isolation, update, numbers, quarantine, positive, cases
Topic 3	safe, completed, health, confirmed, continues, updated, probable, recovery, discharged, last

Text

COVID19 Recovery Rate continues to improve standing at 68.32 with 48,900 **patients discharged** in the **last** 24 **hours**. Actual **case** load for **India** stands at 619,088 today and accounts for 29.64 of total **positive cases**.

9.3.5 CLASSIFICATION OF THE SYNTHESIZED DOCUMENTS

Deep learning has proven to be more efficient because the computational power has considerably increased and because these algorithms can generate more complex functions than other classifiers such as the naïve Bayes (NB), support vector machine (SVM), or random forest.

This research implements an artificial neural network (ANN) with backpropagation to conduct the classification process. ANNs generate nonlinear functions that become more robust through several hidden layers. This process provides more efficient models at the expense of higher computational costs.

The specific parameters used in this study were the following: the number of hidden layers was set to 100, and the activation function used was ReLU.

Once the new documents that avoid secondary ideas were generated, the next step was to classify them into deceptive and truthful texts. Four weighting term methods were used and combined to extract lexical (OHE and TF-IDF) and semantic information (LDA and D2V). In this step, the documents are not divided into sentences but are completely processed using each mapping method.

The five-fold cross-validation technique was used to validate the performance of the classification.

9.4 DATASETS

Two datasets were collected in two languages (English and Arabic) to obtain reliable results in the experiments.

9.4.1 COVID-19 RUMORS

ArCOV19-Rumors (Haouari et al. 2020) is an Arabic COVID-19 Twitter dataset for misinformation detection composed of tweets containing claims from 27 January until the end of April 2020. The dataset consists of 1753 true and 1831 false tweets.

To build the datasets, the authors collected COVID-19 verified claims. Then, for each claim, they found the corresponding relevant tweets. Finally, among the relevant tweets, the authors identify those that either express the claim or negate it. The COVID-related claims were collected from two popular Arabic fact-checking platforms.

For each collected claim, the authors constructed a set of search queries to retrieve all potential tweets. Some tweets that were considered relevant were tweets expressing or negating the claim; tweets having multiple claims including the target claim; tweets expressing advice, questions, or personal opinions about the claim; or even sarcastic tweets about it.

After collecting relevant tweets for each claim, the authors identified those that are expressing or negating the claim; thus, they can propagate the label of the claim to them. Each claim was labeled according to these categories:

- Expressing same claim: If the main claim in the tweet is restating, expressing, or rephrasing the target claim, this tweet then receives the same veracity as the target claim; therefore, it inherits its exact label (truth or deceptive).
- Negating the claim: If the main claim in the tweet is negating or denying the target claim, the veracity of this tweet is then the opposite of the veracity of the target claim.

9.4.2 COVID-19 FAKE NEWS

This dataset (Patwa et al. 2020) consists of 5,600 truthful and 5,100 deceptive reviews. First, the authors collected fake news data from public fact-verification websites and social media. Various web-based resources such as Facebook posts, tweets, news stories, Instagram posts, public statements, press releases, and any other popular media content were used to find fake news content. Second, to collect real claims, the authors crawled tweets from the official and verified Twitter handles of the relevant sources using a Twitter API. The relevant sources are official government accounts, medical institutes, and news channels, among others.

Each tweet was read by a human and was marked as real news if it contained useful information on COVID-19, such as numbers, dates, vaccine progress, government policies, hotspots, etc. The statistics of the dataset and the distribution of data are shown in Table 9.2 and Table 9.3, respectively.

TABLE 9.2

COVID-19 fake news statistics

	Truthful	Deceptive	Combined
Tokens	19,728	22,916	37,503
Avg. tokens per doc	21.65	31.97	27.05

TABLE 9.3
Distribution of documents

Data	Truthful	Deceptive	Total
Training	3,360	3,060	6,420
Validation	1,120	1,020	2,140
Test	1,120	1,020	2,140
Total	5,600	5,100	10,700

9.5 RESULTS

As mentioned above, the main step in this approach is to remove secondary ideas to analyze and generate features only considering the main information in documents. This process involves a clustering representation that considers sentences as objects to find those that could be a centroid. These centroids are the sentences that represent the partitions generated using a GA algorithm. Therefore, each sentence should belong to a certain group after applying the clustering process. Finally, the word distribution of the document provided by LDA helps to select the sentences that will form the new text from the groups or partitions.

To validate our results in the summarization process, a comparison of the performance of the proposed approach with those of other authors was made. The DUC02 dataset was processed, and the quality of the resultant summaries was measured with Rouge (Lin 2004) (a measure that compares ideal summaries with an automatically generated one). The results of this comparison are shown in Table 9.4.

To measure the proximity between objects in vector space, cosine and Euclidean measurements were combined. This combination, as can be seen in Table 9.5, was shown to improve the results in the summarization process (Section 9.3.2).

TABLE 9.4
Comparison of the results of the summarization performance with those of other approaches

Approach	Rouge-1	Rouge-2
This work	0.48681 (1)	0.23334 (1)
UnifiedRank (Wan 2010)	0.48478 (2)	0.21462 (3)
SFR (Vázquez, García-Hernández and Ledeneva 2018)	0.48423 (3)	0.22471 (2)
GA approach (García-Hernández and Ledeneva 2013)	0.48270 (4)	–
COSUM (Alguliyev et al. 2019)	0.46694 (5)	0.12368 (5)
FEOM (Song et al. 2011)	0.46575 (6)	0.12490 (4)
NetSum (Svore, Vanderwende and Burges 2007)	0.44963 (7)	0.11167 (6)
CRF (Shen et al. 2007)	0.44006 (8)	0.10924 (7)

TABLE 9.5

Performance of the proximity measures

Measure	Rouge-1	Rogue-2	Rouge-SU
Cosine	0.47854	0.22615	0.24465
Euclidean	0.47673	0.22332	0.24122
Cosine+Euclidean	**0.48681**	**0.23334**	**0.24954**

The values in this table are represented in terms of unigrams (Rouge-1), bigrams (Rouge-2), and skip-bigrams (Rouge-SU).

Table 9.6 shows the statistics of each corpus analyzed in this study before and after being processed and synthesized. The displayed values are tokens and types, that is, the total number of words and the number of words without repetition in the corpus, respectively; the average tokens per document are also displayed. As the table shows, the number of tokens decreases after the source documents are processed; and thus, the percentage of words that were removed from each corpus is displayed. This percentage varies depending on the LDA selection process detailed in Section 9.3.4.

The next general step is classifying the documents from the selected datasets into deceptive or truthful texts. The corpora analyzed in this study address two languages: Arabic (COVID-19 Rumors) and English (COVID-19 fake news). All corpora were classified using the same configuration detailed above.

Table 9.7 and Table 9.8 show the classification results of the *COVID-19 Rumors* dataset using a neural network and a support vector machine classifier, respectively. Each table shows the methods to generate features and the precision (P), recall (R), and F-measure (F) before and after removing the secondary ideas. As the tables show, the reduction of data in the documents for the Arabic dataset affects the TF-IDF and D2V methods when an SVM is selected (Table 9.8). For example, the F-measure decreases after retaining only the key sentences when applying the TF-IDF method. However, the F-measure increases in all cases when applying the OHE method. In contrast, all the results are improved when the ANN is applied to synthesized documents.

TABLE 9.6

Dataset statistics

Dataset	Full texts				Synthesized texts		
	No. of Docs.	Tokens	Types	Avg. tokens per file	Tokens	Types	Avg. tokens per file
Arabic	3,052	82,792	10,891	27	69,475	8,516	19
English (training/test)	8,560	231,518	20,908	28	160,225	17,849	18

TABLE 9.7

Classification results on Arabic dataset (ANN)

Method	Complete			Summarized		
	P	R	F	P	R	F
OHE	85.01	84.28	84.64	97.47	97.62	97.54
D2V	63.20	63.11	63.15	76.12	76.01	76.06
LDA	64.87	63.44	64.15	65.94	64.98	65.46
TF-IDF	55.27	54.78	55.02	67.30	67.31	67.30
OHE-D2V	86.88	86.32	86.60	96.17	96.25	96.21
OHE-LDA	86.01	85.41	85.71	97.25	97.36	97.30
OHE-TFIDF	85.65	84.40	85.02	96.94	97.09	97.01
ALL	87.05	86.21	86.63	97.07	97.14	97.14

TABLE 9.8

Classification results on Arabic dataset (SVM)

Method	Complete			Summarized		
	P	R	F	P	R	F
OHE	76.04	74.20	75.11	86.14	83.74	84.92
D2V	77.33	77.33	77.33	66.06	65.47	65.76
LDA	27.39	52.33	35.96	63.83	73.67	68.40
TF-IDF	72.38	63.97	67.92	59.73	50.66	54.82
OHE-D2V	77.94	72.71	75.23	81.17	82.22	81.69
OHE-LDA	79.94	79.90	79.92	79.76	84.41	82.02
OHE-TFIDF	78.01	77.85	77.93	83.54	75.43	79.28
ALL	79.61	77.99	78.79	82.38	78.58	80.44

For the proposed framework, the methods that extract lexical information from documents (TF-IDF and OHE) achieve the best results while those that provide semantic information (LDA and D2V), in some cases, are affected by the sentence elimination process. The method that provides the best performance for all datasets is OHE, as Table 9.7 to Table 9.10 show.

Finally, the statistical significance (SS) was calculated by applying a t-test to measure the performance between our approach and those of the other authors (see Table 9.11). A 95% confidence interval was considered, which means that a system will achieve SS if the p-value obtained is below 0.05. According to this table, first, the results obtained for the Arabic language are statistically significant (p-value = 0), and thus our system performs better even when classifying only the key sentences of the documents. Second, the results obtained for the English language do not reach SS (p-value = 0.2412). Therefore, the results of both systems are comparable.

TABLE 9.9

Classification results on English dataset (ANN)

Method	Complete			Summarized		
	P	R	F	P	R	F
OHE	88.79	88.83	88.81	92.39	92.45	92.42
D2V	82.99	83.03	83.01	84.35	84.36	84.35
LDA	65.56	65.33	65.44	68.83	68.57	68.70
TF-IDF	71.84	69.05	70.42	74.10	71.96	73.01
OHE-D2V	89.31	89.35	89.33	92.11	92.16	92.13
OHE-LDA	88.57	88.57	88.57	92.39	92.46	92.42
OHE-TFIDF	89.03	89.05	89.04	92.72	92.78	92.75
ALL	89.31	89.37	89.34	91.88	91.93	91.90

TABLE 9.10

Classification results on English dataset (SVM)

Method	Complete			Summarized		
	P	R	F	P	R	F
OHE	76.52	69.43	72.80	84.22	83.13	83.67
D2V	75.56	54.15	63.09	77.49	77.42	77.45
LDA	27.39	52.33	35.96	65.58	65.60	65.59
TF-IDF	69.42	68.83	69.12	72.32	68.17	70.18
OHE-D2V	74.05	58.92	65.62	82.76	82.75	82.75
OHE-LDA	75.64	69.11	72.23	83.61	82.00	82.80
OHE-TFIDF	76.20	69.06	72.45	85.25	84.67	84.96
ALL	75.46	63.73	69.10	89.37	89.34	89.35

TABLE 9.11

Comparison of the results of the proposed approach with those of other approaches

Method	Lang.	P	R	F	p-value
This work	Arabic	97.25	97.36	97.30	0.000
	English	92.72	92.78	92.75	0.241
ArCOV19-Rumors corpus's authors	Arabic	75.30	74.00	74.00	–
Covid-19 fake news corpus's authors	English	93.48	93.46	93.46	–

9.6 CONCLUSIONS

This study illustrates an approach for COVID-19 fake news detection. Our inference was that the cues to deception stand out in the main ideas of a text. Because the basic unit with a full meaning of a writer's idea is the sentence, the detection of key ideas was addressed by trying to find the most important sentences in each document.

Our research results showed that key sentences are more likely to transmit deception because the classification performance on synthesized documents containing only key sentences can be comparable or even increased with respect to the datasets addressed in this study.

Different methods were proposed for the generation of features to extract information at the lexical and semantic levels, showing that the lexical level was the most appropriate to detect COVID-related fake news for this proposed framework.

In several studies, lexical information provides the most relevant features for detecting deception; however, this study showed that other language levels complement those features and increase the classification performance.

NOTES

1 https://dictionary.cambridge.org/us/dictionary/english/fake-news
2 https://dictionary.cambridge.org/us/dictionary/english/misinformation
3 https://dictionary.cambridge.org/us/dictionary/english/deception
4 https://dictionary.cambridge.org/us/dictionary/english/lie

REFERENCES

Alguliyev, R. M., R. M. Aliguliyev, N. R. Isazade, A. Abdi, and N. Idris. 2019. COSUM: Text summarization based on clustering and optimization. *Expert Systems* (Wiley Online Library) 36, No. 1: e12340.

Bang, Y., E. Ishii, S. Cahyawijaya, Z. Ji, and P. Fung. 2021. Model Generalization on COVID-19 Fake News Detection. *arXiv preprint arXiv:2101.03841*

Blei, D. M., A. Y. Ng, and M. I. Jordan. 2003. Latent dirichlet allocation. *The Journal of Machine Learning Research* (JMLR. org) 3: 993–1022.

Felber, T. 2021. Constraint 2021: Machine Learning Models for COVID-19 Fake News Detection Shared Task. *arXiv preprint arXiv:2101.03717*

Feng, S., R. Banerjee, and Y. Choi. 2012. Syntactic stylometry for deception detection. Vol. 2, In *Proceedings of the 50th Annual Meeting of the Association for Computational Linguistics.* Association for Computational Linguistics, 171–175.

Fornaciari, T., and M. Poesio. 2014. Identifying fake Amazon reviews as learning from crowds. In *Proceedings of the 14th Conference of the European Chapter of the Association for Computational Linguistics.* Association for Computational Linguistics, 279–287.

Fusilier, D. H., M. Montes-y-Gómez, P. Rosso, and R. G. Cabrera. 2015. Detection of opinion spam with character n-grams. In *International Conference on Intelligent Text Processing and Computational Linguistics*, 285–294. Springer.

García-Hernández, R. A., and Y. Ledeneva. 2013. Single extractive text summarization based on a genetic algorithm. In *Mexican Conference on Pattern Recognition*, 374–383. Springer.

Gu, X., P. P. Angelov, D. Kangin, and J. C. Principe. 2017. A new type of distance metric and its use for clustering. *Evolving Systems* (Springer) 8, no. 3: 167–177.

Hamid, A., et al. 2020. Fake News Detection in Social Media using Graph Neural Networks and NLP Techniques: A COVID-19 Use-case. *arXiv preprint arXiv:2012.07517*

Haouari, F., M. Hasanain, R. Suwaileh, and T. Elsayed. 2020. ArCOV19-Rumors: Arabic COVID-19 Twitter Dataset for Misinformation Detection. *arXiv preprint arXiv:2010.08768*

Hernández-Castañeda, N., R. A. García-Hernández, Y. Ledeneva, and Á. Hernández-Castañeda. 2020. Evolutionary automatic text summarization using cluster validation indexes. *Computación y Sistemas* 24, no. 2: 583–595.

Kar, D., M. Bhardwaj, S. Samanta, and A. P. Azad. 2020. No Rumours Please! A Multi-Indic-Lingual Approach for COVID Fake-Tweet Detection. *arXiv preprint arXiv:2010.06906*

Le, Q., and T. Mikolov. 2014. Distributed representations of sentences and documents. *International Conference on Machine Learning* (PMLR), 1188–1196.

Lin, C. 2004. Rouge: A package for automatic evaluation of summaries. *Text Summarization Branches Out*. Vol. 1. Association for Computational Linguistics, 74–81.

Liu, Y., Z. Li, H. Xiong, X. Gao, and J. Wu. 2010. Understanding of internal clustering validation measures. In *2010 IEEE International Conference on Data Mining*, 911–916. IEEE.

Newman, M. L., J. W. Pennebaker, D. S. Berry, and J. M. Richards. 2003. Lying words: Predicting deception from linguistic styles. *Personality and Social Psychology Bulletin* (Sage Publications) 29, no. 5: 665–675.

Patwa, P., et al. 2020. Fighting an infodemic: Covid-19 fake news dataset. *arXiv preprint arXiv:2011.03327*

Polage, D. 2017. The effect of telling lies on belief in the truth. *Europe's Journal of Psychology* (PsychOpen) 13, no. 4: 633.

Rendón, E., I. Abundez, A. Arizmendi, and E. M. Quiroz. 2011. Internal versus external cluster validation indexes. *International Journal of Computers and Communications* 5, no. 1: 27–34.

Rosso, P., and L. C. Cagnina. 2017. Deception detection and opinion spam. Cambria, E., Das, D., Bandyopadhyay, S., Feraco, A. (eds). Vol. 5, In *A Practical Guide to Sentiment Analysis*, 155–171. Springer, Cham.

Rosso, P., F. Rangel, I. H. Farías, L. Cagnina, W. Zaghouani, and A. Charfi. 2018. A survey on author profiling, deception, and irony detection for the arabic language. *Language and Linguistics Compass* (Wiley Online Library) 12, no. 4: e12275.

Schelleman-Offermans, K., and H. Merckelbach. 2010. Fantasy proneness as a confounder of verbal lie detection tools. *Journal of Investigative Psychology and Offender Profiling* (Wiley Online Library) 7, no. 3: 247–260.

Shahi, G. K., and D. Nandini. 2020. Fake Covid–A Multilingual Cross-domain Fact Check News Dataset for COVID-19. *arXiv preprint arXiv:2006.11343*

Shen, D., J. Sun, H. Li, Q. Yang, and Z. Chen. 2007. Document summarization using conditional random fields. Vol. 7, In *Proceedings of the 20th international joint conference on Artifical intelligence*. Morgan Kaufmann Publishers Inc., 2862–2867.

Song, W., L. C. Choi, S. C. Park, and X. F. Ding. 2011. Fuzzy evolutionary optimization modeling and its applications to unsupervised categorization and extractive summarization. *Expert Systems with Applications* (Elsevier) 38, no. 8: 9112–9121.

Svore, K., L. Vanderwende, and C. Burges. 2007. Enhancing single-document summarization by combining RankNet and third-party sources. In *Proceedings of the 2007 Joint Conference on Empirical Methods in Natural Language Processing and Computational Natural Language Learning (EMNLP-CoNLL)*. Association for Computational Linguistics, 448–457.

Toma, C. L., and J. T. Hancock. 2012. What lies beneath: The linguistic traces of deception in online dating profiles. *Journal of Communication* (Oxford University Press) 62, no. 1: 78–97.

Vázquez, E., R. A. García-Hernández, and Y. Ledeneva. 2018. Sentence features relevance for extractive text summarization using genetic algorithms. *Journal of Intelligent & Fuzzy Systems* (IOS Press) 35, no. 1: 353–365.

Wan, X. 2010. Towards a unified approach to simultaneous single-document and multi-document summarizations. In *Proceedings of the 23rd international conference on computational linguistics (COLING 2010)*. CONLING 2010 Organizing Committee, 1137–1145.

Xu, R., and D. Wunsch. 2008. *Clustering*. Vol. 10. John Wiley & Sons.

Xu, Q., and H. Zhao. 2012. Using deep linguistic features for finding deceptive opinion spam. In *Proceedings of COLING 2012: Posters*. COLING 2012 Organizing Committee, 1341–1350.

10 Employing Computational Linguistics to Improve Patient-Provider Secure Email Exchange: The ECLIPPSE Study

Renu Balyan
Department of Mathematics, Computer & Information
Sciences, State University of New York at Old Westbury,
New York, USA

Danielle S. McNamara
Department of Psychology, Arizona State University, Tempe,
Arizona, USA

Scott A. Crossley
Department of Applied Linguistics/ESL, College of Arts and
Sciences, Georgia State University, Atlanta, GA, USA

William Brown III
Center for AIDS Prevention Studies, University of California,
San Francisco, San Francisco, CA, USA

UCSF Health Communications Research Program, Center for
Vulnerable Populations, Zuckerberg San Francisco General
Hospital and Trauma Center, San Francisco, California, USA

Andrew J. Karter
UCSF Division of General Internal Medicine at Zuckerberg
San Francisco General Hospital and Trauma Center,
San Francisco, California, USA

Division of Research, Kaiser Permanente Northern
California, Oakland, California, USA

DOI: 10.1201/9781003138013-10

Dean Schillinger
UCSF Division of General Internal Medicine at Zuckerberg
San Francisco General Hospital and Trauma Center,
San Francisco, California, USA

Division of Research, Kaiser Permanente Northern
California, Oakland, California, USA

UCSF Health Communications Research Program, Center for
Vulnerable Populations, Zuckerberg San Francisco General
Hospital and Trauma Center, San Francisco, California, USA

CONTENTS

10.1   INTRODUCTION

As per the Centers for Disease Control and Prevention (2020), an estimated 34.2 million people in the United States (10.5% of the U.S. population) in 2018 were suffering from diabetes mellitus (DM). Self-managing DM, like with most chronic conditions, can be complex and requires frequent communication in between visits with healthcare providers. *"Health literacy (HL) –defined as a patient's ability to obtain, process, comprehend, communicate and act on basic health information"* (Protection and Act 2010; Schillinger et al. 2017) – plays a critical role for DM patients. Limited HL results in poor health outcomes (Schillinger et al. 2002; Schillinger et al. 2004; Sarkar et al. 2010), and contributes to preventable suffering, excess healthcare costs, and rapid physical functions decline (Smith et al. 2015). Online patient portals (e.g., Kaiser Permanente Northern California: KPNC patient portal, kp.org) are widely being used to provide communication support to patients and providers via SMs. Patients accessing such portals exhibit better (a) medication adherence (Sarkar et al. 2014; Lyles et al. 2016), (b) healthcare utilization (Reed et al. 2013), and (c) glycemic (blood sugar) control (Reed et al. 2012; Harris et al. 2013). However, the effectiveness of such web-based communication measures can be influenced by the HL of a patient. The overarching goal of the ECLIPPSE project was to identify ways to harness such online communication to reduce health disparities related to HL.

10.2   THE ECLIPPSE PROJECT

The ECLIPPSE (Employing Computational Linguistics to Improve Patient-Provider Secure Emails exchange) is a National Library of Medicine (NLM) funded project that aimed to: a) develop and validate an automated LP by assessing the linguistic features of secure messages (SMs) generated by patients and sent to their primary care physicians (Balyan et al. 2019; Crossley et al. 2021; Schillinger et al. 2020); b) demonstrate the validity of a patients' LP by examining associations with measures of patients' HL, their reports of their providers' communication, and several diabetes outcomes (Balyan et al. 2019; Crossley et al. 2021; Schillinger et al. 2020); c) develop and validate an automated CP using the linguistic features of the SMs written by primary care physicians to their DM patients (Crossley et al. 2020); d) determine the prevalence of LP-CP discordance (if any exists) across physician-patient dyads and explore the outcomes; and e) create and evaluate an automated LP and CP-based patient portal prototype that provides providers feedback in real-time to reduce the their SMs complexity and better accommodate patients' HL while they are writing SMs.

Two recent studies carried out as part of the ECLIPPSE project assessed the NLP tools' capability to classify HL of a patient (Balyan et al. 2019; Crossley et al. 2021). Both the studies used NLP tools on the corpus created by combining SMs sent by patients to their clinicians for developing patients' self-reported HL (obtained via survey) and expert- rated HL ML models. Their models reported results similar to previous HL research: patients classified by the model as having limited HL had higher probability of being older, belonging to a minority group, and attained limited education. In addition, these patients had higher rates of hypoglycemia (an adverse drug event), comorbidities, healthcare utilization, worse medication adherence, and poor blood sugar control. Crossley et al. (2020) additionally leveraged expert ratings of the SMs to examine the linguistic complexity of the SMs sent by physicians to their patients so as to develop a physician CP. A depiction of the workflow for the overall process is shown in Figure 10.1.

10.3   IMPORTANCE OF CHARACTERIZING PATIENT HEALTH LITERACY

HL is representative of procative, interactive and effective set of skills and communication behaviors (Sudore et al. 2009) between patients and clinicians including skills related to literacy, verbal (speaking and listening), numeracy, and health and healthcare-related digital skills. Research shows that poor communication is an important arbitrator in health outcomes and limited HL relationships, and that improving communication can mitigate health disparities associated with HL (Schillinger et al. 2004; Schillinger et al. 2009).

In the United States, older populations and socio-economically disadvantaged, including those of low income or education, some minority groups, and English-speaking immigrants are more likely to have limited HL (Kirsch et al. 2002; Institute of Medicine 2004; Smith et al. 2015). Limited HL has been found to be associated with poor health and mortality across populations and medical conditions (Sudore et al. 2006). In the diabetes context, limited HL is associated with a higher prevalence of DM, worse glycemic control, severe hypoglycemia (Sarkar et al. 2010), worse medication adherence (Karter et al. 2009), and higher rates of complications (Schillinger et al. 2002). Therefore, limited HL represents a costly and critical public and clinical health problem that is potentially remediable (Institute of Medicine 2004; Bailey et al. 2014).

One solution to this problem is to increase the quantity and quality of patient-physician electronic communications. Patients accessing such measures are more likely to adhere to prescribed regimens, achieve better outcomes, and have favorable patterns for healthcare utilization (Zhou et al. 2010). Technology platforms when developed in collaboration with and for the limited HL patients, and are accessible to them can indeed disproportionately improve health outcomes and support those patients that have the greatest communication needs (Schillinger 2007; Schillinger et al. 2008).

Patient portals as a source of communication via SMs are gaining greater importance, and enabling asynchronous and between-visit communications. Therefore, patients must achieve some level of digital communicative HL skills to take advantage of these portals. SM exchange is more relevant for chronic illnesses such as DM patients, because of their frequent inter-visit needs for communication. However, patients having limited HL may find it difficult to message their provider or understand their instructions or responses (Sarkar et al. 2010). Therefore, a patients' ability to effectively process (understand and act on) and write SMs is an inherent skill set to patient HL.

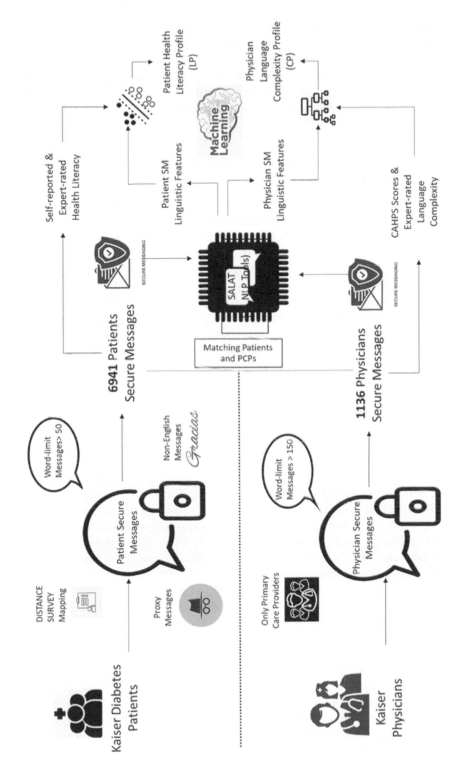

FIGURE 10.1 ECLIPPSE project overall workflow.

Current measurement tools for HL present numerous challenges for administrating and scalability because they are time-consuming and need in-person administration. A more efficient approach to identify limited HL using NLP and ML techniques could improve quality, inform care management initiatives (DeWalt et al. 2011; Karter et al. 2015), reduce disparities related to communication (Seligman et al. 2005), and target and tailor strategies related to population management (Brach et al. 2012).

10.4 IMPORTANCE OF CHARACTERIZING PHYSICIAN WRITING COMPLEXITY

In parallel, physicians need to develop communication skills and adopt the behaviors so as to tailor their written communications to match the needs of their patients. Providers must engage with patients in a way that provides actionable and meaningful information, and promotes shared meaning by providing comprehensible support (Brach, Dreyer, and Schillinger 2014). Physicians rarely employ recommended communication strategies for patients with limited HL (Schillinger et al. 2003), and frequently use clinical jargon (Castro et al. 2007), which can be incomprehensible in particular to patients having limited HL. Most physicians are unaware of their patients' HL, yet appear to be responsive after receiving this information, and utilize more recommended strategies (Seligman et al. 2005). Furthermore, research has demonstrated that HL-related disparities may be diminished by reducing the health communications literacy demands directed towards patients (DeWalt et al. 2009; Schillinger et al. 2009; Baker et al. 2011).

Not much is known about the readability of physicians' SMs. Use of classic readability formulas for assessing medical and health-related text has frequently indicated that many adults find the texts difficult to comprehend as these are written at higher levels (Berland et al. 2001; Kusec, Brborovic, and Schillinger 2003; Boulos 2005; Kandula and Zeng-Treitler 2008; Walsh and Volsko 2008; Hill-Briggs, Schumann, and Dike, 2012; McAndie, Gilchrist, and Ahamat 2016; Kugar et al. 2017; Schumaier et al. 2018). Approximately 89 million people in the US have been estimated to not understand and process most of the available health-related materials (Kindig, Panzer, and Nielsen-Bohlman 2004). The American Medical Association and the National Institutes of Health suggest that patient-oriented health texts should be written at lower grade levels (6th–8th grades; Badarudeen and Sabharwal 2010; Kugar et al. 2017).

Several problems have been reported in terms of measuring the readability, predictability, and reliability of physicians' SMs, when using existing classic readability methods (Meade, Byrd, and Lee 1989; Friedman and Hoffman-Goetz 2006; Wang et al. 2013; Wu et al. 2013; Barton et al. 2016; Zheng and Yu 2018). There are no efficient tools available for assessing the complexity of SMs written by the physician(s) to their patient(s) (Grabeel et al. 2018; Crossley et al. 2020). Therefore, development of a robust and reliable measure for physicians' SM linguistic complexity when used with a patient HL measure, could enable identification and determination of physician-patient linguistic discordance, ascertain its proximal communication consequences, and clinical outcomes. Furthermore, developing such a measure would assist and enable health systems in identifying physicians who might be in need of and benefit from additional communication-related training and support (Kim et al. 2007; Oliffe et al. 2019).

10.5 NLP APPROACHES: HEALTH LITERACY AND TEXT READABILITY

While NLP has long been used in the medical domain for numerous applications including semantic lexicon development, clinical narrative representation, and text quality assessment (Johnson 1999), only a few studies have examined the use of NLP in HL for investigating the readability of medical texts and use of NLP features to predict readability. To our knowledge, no NLP research has assessed physicians' oral or written communications to their patients.

10.5.1 NLP AND HEALTH LITERACY

A recent HL research measures review has found around 200 HL measures, out of these 52% measures required paper and pencil, and 12% required more than 15 minutes to administer. None of these studies has assessed the SMs written by patients' to their physicians to measure a patients' communicative HL. In general literacy studies have shown that writing skill (i.e., linguistic production) is strongly correlated with reading skill (i.e., linguistic comprehension), thus providing a strong reason to harness patients' SMs for assessing communicative HL.

As a result, employing NLP tools could be one approach to identify patients' HL for assessing the writing skill of a patient. NLP approaches can efficiently analyze data in huge amounts that is tedious and time-intensive for humans to accomplish. An automated HL measure using SMs may help overcome obstacles and challenges (e.g., scaling for larger patient populations) reported previously while conducting interviews or questionnaires for measuring HL. An automated NLP-based HL classification measure would provide an efficient means to automatically identify at-risk patients having limited HL, and requiring interventions.

10.5.2 NLP AND TEXT READABILITY

Most of the research assessing medical text readability has used readability formulas such as the Dale-Chall – created for students in grade 4 or above, the scores are based on number of unfamiliar words in the text (Chall and Dale 1995); Flesch-Kincaid Grade Level (FKGL) – uses sentence length and syllables in a word, weighing longer sentences more than words (Kincaid et al. 1975); Flesch Reading Ease (FRE) computes a score between 1 and 100 but weights longer words more as compared to long sentences (Flesch 1948); Simple Measure of Gobbledygook (SMOG) – used in healthcare for medical writing and considers the percentage of words having three or more than three syllables (McLaughlin 1969), Frequency of Gobbledygook – takes into consideration the percentage of long words and the sentence length (Gunning 1952), and Fry – used across a range of sectors and also uses sentence and syllables for computing the scores (Fry 1968). However, performance, validity, and effectiveness of these formulas in medical and other domains (Wang et al. 2013; Wu et al. 2013; Zheng and Yu 2017) have been strongly questioned. One possible explanation for the ineffectiveness of these readability formulas is absence of a strong mapping between their linguistic features (i.e., word and sentence length) and the linguistic constructs that are predictive of reading comprehension (Bruce, Rubin, and Starr 1981; Davison and Kantor 1982; Rubin 1985; Bruce and Rubin 1988; Graesser, McNamara, and Kulikowich 2011; Smith 2012; Balyan, McCarthy, and McNamara 2018; McNamara et al. 2019; Balyan, McCarthy, and McNamara 2020). In response, studies by readability researchers across a variety of domains have investigated and successfully shown NLP features capability to predict readability for medical texts (Kim et al. 2007; Zeng–Treitler et al. 2012; Wu et al. 2013; Zheng and Yu 2018).

In the medical domain, researchers have augmented the basic features with advanced NLP measures including lexical sophistication/diversity (Gemoets et al. 2004), word familiarity (Zheng and Yu 2018), syntactic, semantic, and cohesion features (Gemoets et al. 2004; Kim et al. 2007; Zeng–Treitler et al. 2012), and part of speech (PoS) tags (Kim et al. 2007; Zeng–Treitler et al. 2012). These new readability models that use NLP features perform better than the classic readability formulas in the medical domain (Kim et al. 2007; Zeng–Treitler et al. 2012) as well as outside of the medical domain (Crossley, Greenfield, and McNamara 2008; Pitler and Nenkova 2008; De Clercq et al. 2014; Crossley et al. 2017).

No research has focused on physicians' CP. However, automatically assessing CP is an important component of understanding how physicians interact with patients and whether or not patients can understand and act on suggestions made by the physician. Adjusting messages written to patients is a cumbersome task, especially when existing readability formulas are not efficient and effective, there is no certainty about why medical texts are difficult to comprehend, and physician training is lacking. Even then, there is little research that has examined physicians' written communication readability for messages written to patients.

10.6 TOOLS FOR LINGUISTIC FEATURES EXTRACTION

A Suite of Automatic Linguistic Analysis Tools (SALAT[1]) was used to extract linguistic features from the patient-physician SMs that measure different aspects of language, including text level information, lexical sophistication, syntactic complexity, and text cohesion. These NLP tools in turn use several other resources including Stanford Parser (De Marneffe, MacCartney, and Manning 2006), Wordnet (Miller 1995), CELEX word frequency database (Baayen, Piepenbrock, and Gulikers 1996), a psycholinguistic database (Coltheart 1981), a corpus (BNC; BNC Consortium 2007), and medical corpora including HIMERA (Thompson et al. 2016), i2b2[2] (Uzuner, Luo, and Szolovits 2007; Uzuner et al. 2008; Uzuner 2009; Uzuner, Solti, and Cadag 2010).

The SALAT consists of tools including *TAACO* (Crossley, Kyle, and McNamara 2016) – incorporates indices for local and global text cohesion analysis at both word and sentence levels; *TAALES* (Kyle and Crossley 2015) – constitutes tools for automatically assessing lexical sophistication, measures of concreteness of words, their familiarity and meaningfulness (Kyle, Crossley, and Berger 2018); *TAASSC* (Kyle 2016; Crossley, Kyle, and McNamara 2017) – measures clausal (31) and phrasal (132) syntactic complexity indices, 190 indices of syntax sophistication, and 14 indices from the syntactic complexity analyzer by Lu (Lu 2010); and *SEANCE* (Crossley et al. 2017) – a sentiment analysis tool using a PoS tagger, and a number of dictionaries for sentiment, cognition, and social positioning, along with a negation feature. In addition to the SALAT, a tool for writing assessment (*WAT*; Crossley, Roscoe, and McNamara 2013) including indices specific to writing (text structure, cohesion, lexical sophistication, syntactic complexity, and rhetorical features) was also used (McNamara, Crossley, and Roscoe 2013).

The tools discussed in this section including TAALES and TAACO have been used for essays as well as unstructured data sets including, but not limited to, first and second language speech samples, children's e-mails, forum posts, early childhood writing, and beginning level L2 writing. Therefore, we decided to use these tools for SMs exchanged between physicians and patients.

10.7 MACHINE LEARNING APPROACHES

Several supervised machine learning classification models were used while building the patient LPs and physician CPs. A brief description of the algorithms is provided in Table 10.1 (Hastie, Tibshirani, and Friedman 2009; James et al. 2013; Balyan, McCarthy, and McNamara 2017). The use of linguistic and semantic indices was motivated by theoretical models of literacy and text complexity.

Our objective is to develop models of literacy and complexity that are linked to theoretical models of discourse and capable of driving feedback. While neural methods are gaining more focus these days and are indeed powerful, such methods result in less interpretable (explainable) and oftentimes highly-resource intensive models. As such, the decision to not use neural methods was driven by both the theoretical and applied requirements of this project. Likewise, the continued use of linguistic and semantic features in NLP models is crucial to the advancement of this field, as well as our understanding of language.

10.8 DEVELOPING AND VALIDATING PATIENTS' LITERACY PROFILES

Current direct measures of HL make widespread classification of patient HL challenging because they are time demanding and can be invasive. Hitherto, limited HL patients identification has proven laborious and infeasible to scale (DeWalt et al. 2011). Therefore, the "Literacy Profiles" were developed and validated as patients' HL automated measures for facilitating an economical, non-invasive, and scalable characterization of HL.

10.8.1 DATA SOURCE AND PARTICIPANTS

The study used data from the KPNC (a fully integrated health system with ~4.4 million patients) Diabetes Registry consisting more than a million SMs written

TABLE 10.1
Machine learning models' brief description

Method Type	ML Algorithm	Description
Single Methods	Naïve Bayes	Naïve Bayes is a probabilistic method based on the posterior probability of Bayes' theorem. It makes an assumption about the features of independence (McCallum and Nigam 1998).
	Decision Trees	The Decision Trees represents the learned function as a set of if-then rules. The feature for a node is determined by information gain or gini index statistical properties (Mitchell 1997).
	Artificial Neural Networks	Artificial Neural Networks (ANN) model the human nervous system processing capabilities for information. These models use algorithms such as back-propagation for updating weights based on the feedback and are self-learning (Zhang 2000; Rojas 2013).
	Linear Discriminant Analysis	Linear Discriminant Analysis (LDA) is a "supervised" linear transformation technique that computes "linear discriminants" representing the axes that maximize the separation between multiple classes and is commonly used for reducing data dimensions (Fisher 1936).
	Support Vector Machine	Support Vector Machine (SVM) classify data into different classes by constructing a hyperplane. SVMs are among the best supervised learning algorithms that are efficient for high-dimensional data (Dumais et al. 1998; Joachims 1998).
Ensemble methods	Random Forests	Random Forests overcome the decision trees "overfitting" problem and construct multiple decision trees during the training phase implementing majority voting for classification (Smola and Schölkopf 1998; Schapire and Singer 2000).
	Bagging	Bootstrap Aggregation (Bagging) considers multiple classifiers, is a meta-algorithm creating training set bootstrap samples (sampling with replacement), and uses majority voting for classification (Breiman 1996).
	Boosting	Boosting like Bagging is a meta-algorithm. It iteratively trains multiple weak learners to build an ensemble and uses misclassified instances from the previous models. The final result is based on weighted sum of all classifier results (Krogh and Vedelsby 1995).
	Stacking	Stacked generalization (Stacking) creates an ensemble by combining multiple classifiers generated on a single data set. It creates a set of base classifiers and combines their outputs to train a meta-level classifier (Wolpert 1992).

by >150, 000 diabetes patients and >9,000 primary care physicians. The patients from KPNC registry who participated and completed a 2005–2006 DISTANCE survey (Diabetes Study of Northern California) and responded to the items related to self-reported HL (N = 14,357; Chew et al. 2008; Moffet et al. 2009; Ratanawongsa et al. 2013) were used. The survey details can be seen in Figure 10.2. Further details of the DISTANCE Study have been reported previously (Moffet et al. 2009). The data was collected using questionnaires completed online, via paper and pencil or on the telephone with a response rate of 62%. DISTANCE surveyed diabetes patients with average age 56.8 years (±10); and who were male (54.3%); Latino (18.4%), African American (16.9%), Caucasian (22.8%), Filipino (11.9), Asian (Chinese/Japanese; 11.4%), South Asian/ Native American/Pacific Islander/ /Eskimo (7.5%), and multi-racial (11.0%).

The original corpus included all (N = 1,050,577) diabetes patients-clinicians exchanged SMs from KPNC's patient portal from January 1, 2006 through December 31, 2015. However, SMs that patients exchanged with only their primary care physicians (PCPs) were included in the analyses described in the study. The SMs were filtered for the patients whose DISTANCE survey data was missing. In addition, non-English SMs (a less than 1% messages were removed) and those SMs not written by the patient but the proxy caregivers (determined by a proxy check-box in the KP.org or validated by an NLP algorithm; Semere et al. 2019) were also excluded. The final ECLIPPSE dataset for the patient HL measure constitutes 283,216 SMs written by 6,941 patients to their PCPs. All of the SMs for each patient were aggregated into a single file. The patients having less than 50 words in the aggregated SMs file were excluded. Previous research in NLP text in learning analytics was the determining factor for the 50-word exclusion threshold (Crossley et al. 2016; Crossley and Kostyuk 2017). All of the survey and SM data were confidential and stored and analyzed on secure servers behind KPNC firewalls that precluded copying and downloading, and maintained data security. Therefore, we are not in a position to provide any examples because the data used for this study contains protected health information (PHI) and access is protected by the KPNC Institutional Review Board (IRB).

10.8.2 HEALTH LITERACY MEASURES

Two LP measures based on NLP and ML were developed. In the first LP, a validated HL scale that contained three items measuring self-efficacy in specific HL competencies from the DISTANCE survey was included. The items were self-reporetd by patients using a 5-point Likert scale (1: "Always" and 5: "Never"; Sarkar et al. 2011). The items measured patients' confidence in filling medical forms, and patients' understanding of written medical information, and frequency of help needed while reading health documents. The new self-reported HL variable used in the study was created by computing average scores across these HL items. These scores were dichotomized to indicate limited and adequate HL (Balyan et al. 2019). The threshold considered for dichotomization was consistent with scale used in previously employed studies (Chew et al. 2008; Sarkar et al. 2008; Sarkar et al. 2011).

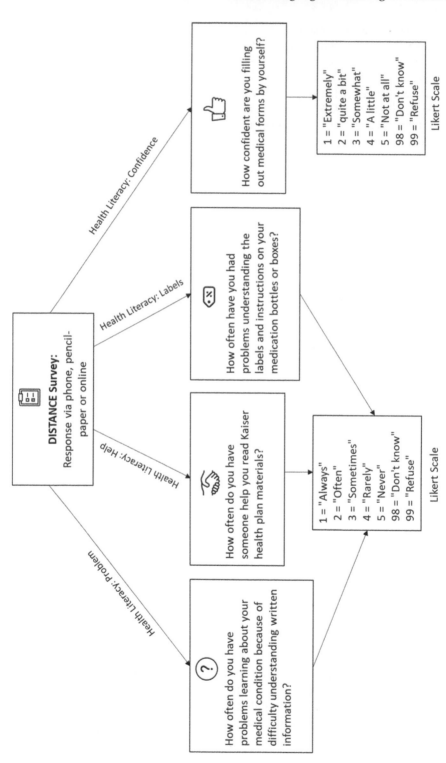

FIGURE 10.2 The DISTANCE survey health literacy-related items.

Because of concerns related to possible mismeasurement using self-report, a second LP was developed, in which HL was measured based on the quality of patients' SMs, using ratings by experts. These ratings used a subset of aggregated SMs written by 512 patients from the DISTANCE survey (Crossley et al. 2021). A scoring rubric to assess a patients' perceived HL based on their SMs was adapted from a rubric used for testing high school seniors writing abilities (Crossley, Kyle, and McNamara 2015). The patients' written English proficiency, organization, health vocabulary accuracy, and patients' intent towards their physician were assessed using a scale ranging from 1: low HL to 6: high HL (Crossley et al. 2021).

10.8.3 Creating Literacy Profiles

Using NLP and ML techniques, five separate LP prototypes (Schillinger et al. 2020), each varying in their sophistication and linguistic features, were developed for classifying patients' HL (self-reported; Balyan et al. 2019, and expert-rated; Crossley et al. 2021). Table 10.2 summarizes the LPs and the linguistic indices used while creating each LP.

10.8.4 Evaluating Literacy Profiles

Associations between these LPs' classifications of HL and patients' education, race/ethnicity, and age were then explored. Because associations are also known to exist between HL and communication between the patient and the provider (Schillinger et al. 2003, 2004; Castro et al. 2007; Sarkar et al. 2008), relationships between the LPs and patients' reports of physician communication from the Consumer Assessment of Healthcare Provider and Systems (CAHPS) survey (Schillinger et al. 2004), reported via the DISTANCE survey, were also examined. Associations between the LPs and diabetes-related outcomes, including adherence to medications based on continuous medication gaps (CMG; Steiner et al. 1988; Steiner and Prochazka 1997), hemoglobin A1c (HbA1c), hypoglycemia (Sarkar et al. 2010), and comorbidities were also examined. Finally, relations between each LP and healthcare utilization (outpatient, emergency room, and hospitalization utilization) were explored.

The findings are summarized in Table 10.3. The results in Table 10.3 are not a result of correlations but associations between the predictions made by the five literacy profiles and the patient health data available in their EHRs. Since the focus of the chapter was to show that NLP is capable of predicting the health literacy of a patient based on the SMs written by the patient and the LP followed the patterns similar to prior research in health domain, we did not include the health outcomes of the patients but only showed which LP was able to significantly predict or followed patterns known in the health domain. These results have been previously published (Schillinger et al. 2020).

In addition, the impact of race and ethnicity on the accuracy of models is indeed an extremely important topic and concern. We have also demonstrated that the LP models were not impacted by race (Schillinger et al. 2021).

TABLE 10.2

Linguistic indices and the literacy profiles* (Schillinger et al. 2020)

Literacy Profile	Linguistic Indices	Literacy Profiles Description
Flesch Kincaid (LP_FK)	*Readability:* length of words and sentences	Flesch-Kincaid, being the simplest, most widely available, and most commonly used readability formulas in the medical domain, was used as a baseline measure (Paasche-Orlow, Taylor and Brancati 2003; Wilson 2009; Piñero-López et al. 2016; Jindal and MacDermid 2017; Munsour et al. 2017; Zheng and Yu 2018).
Lexical Diversity (LP_LD)	*Lexical Diversity:* D-based words variety	Lexical diversity (LD), another commonly used method in the linguistics domain for assessing writing proficiency (Malvern et al. 2004; McCarthy 2005) captures cohesion and lexical richness of text. Both these features are consistent predictors of writing quality and text sophistication (McNamara et al. 2014; McNamara et al. 2015). As a result, lexical diversity model in addition to the Flesch-Kincaid model was used as a second baseline.
Writing Quality (LP_WQ)	*Word Frequency:* reference corpus words frequency *Syntactic Complexity:* words count before the main verb within a sentence *Lexical Diversity:* words variety based on MTLD.	A model (McNamara, Crossley, and McCarthy 2010) previously validated, derived from word frequency and syntactic complexity in addition to the LD (McCarthy 2005; McNamara et al. 2014) was used to classify patients' SMs based on the writing quality as low or high HL. The model was developed to test the significance of adding linguistic features on top of the existing LD feature.
Self-Reported (LP_SR)	*Concreteness:* word concreteness degree *Lexical diversity:* MTLD and D-based words variety. *Present tense:* incidence *Determiners:* incidence *Adjectives:* incidence *Function words:* incidence	The self-reported HL profile was created using 185 linguistic features extracted from the patients' written SMs. Some of the key features used in this LP are explained in this table. The motivation, rationale, development, and experimental details for LP_SR have been reported previously (Balyan et al. 2019).
Expert-Rated (LP_Exp)	*Age of Exposure:* estimated age of a word appears in a child's vocabulary *Lexical decision response time:* time taken by humans in judging characters *Attested lemmas:* count per verb argument construction *Determiner per nominal phrase:* determiner count in a noun phrase (NP) *Dependents per nominal subject:* subject structural dependent count in an NP *Number of associations:* with each word	The expert-rated LP was developed using eight linguistic indices to predict expert human ratings of HL. The LP was developed using a sampled subset of 512 expert-rated SMs. Additional details related to the LP_Exp development and experimental design have been previously reported (Crossley et al. 2021).

TABLE 10.3
Results for health literacy correlates vs literacy profiles

Literacy Profile / Health Literacy Correlate		FK	LD	WQ	SR	Exp	Summary
Sociodemographic	Race	√	√	√	√	√	All the LPs classified limited HL to be associated with non-white race. Three LPs (LP_LD, LP_SR, and LP_Exp) were associated with lower education, with LP_SR and LP_Exp observing the strongest effects. Only one LP (LP_SR) was associated with older patient age.
	Education	X	√	X	√	√	
	Age	X	X	X	√	X	
Provider Communication		X	X	X	√	√	The patients predicted as having limited HL by two LPs (LP_SR and LP_Exp) were more likely to rate communications with their health care providers as "poor" based on the CAHPS item. The findings for LP_SR were found to be somewhat more robust.
Health Outcomes	CMG	√	X	X	√	√	Limited HL classified by LP_FK, LP_SR, and LP_Exp was associated with poor medication adherence, greater comorbidity, and serious hypoglycemia. Poor medication adherence, and optimal and poor diabetes control was most significantly associated with LP_FK and LP_Exp.
	Hypoglycemia	√	X	X	√	√	
	Diabetes control	√	X	X	X	√	
Healthcare utilization	Outpatient visits	X	X	X	√	X	The only model capable of associating limited HL with higher rates of hospitalizations and outpatient visits was LP_SR. Both LP_SR and LP_Exp observed higher annual emergency room utilization rates for limited HL, but the differences related to HL were more robust for LP_SR.
	Hospitalization	X	X	X	√	X	
	Emergency room utilization	X	X	X	√	√	

Notes: FK: Flesch Kincaid; LD: Lexical Diversity; WQ: Writing Quality; SR: Self-Reported; Exp: Expert-Rated.

10.9 DEVELOPING AND VALIDATING PHYSICIAN COMPLEXITY PROFILES

Diabetes patients with ongoing self-management needs require counseling and guidance, and often use SMs to discuss issues such as test results, medication concerns, emerging symptoms, requests for appointment and referral, and responses to treatment (Friedman and Hoffman-Goetz 2006). Physician, in turn, must respond to these concerns. Insofar as shared understanding is a critical objective of communication exchange, identifying the linguistic features that make physicians' SMs more or less difficult to understand needs to be pursued. Hence, physicians' SMs to their patients were harnessed as a means to develop an automated readability formula based on advanced linguistic features associated with text complexity of the SMs, which was called the physician CP. To do so, linguistic features and the expert ratings of physician text complexity relationships were examined. Here again, as a reference, the newly developed text complexity profile performance was compared to the Flesch-Kincaid (Kincaid et al. 1975).

10.9.1 DATA SOURCE AND PARTICIPANTS

The data used in this study was derived from SMs drawn from the ECLIPPSE Study (already discussed in section 10.8.1) and exchanged between January 1, 2006 – December 31, 2015. A total of 1,136 primary care physicians were found to have sent SMs to patients. Based on prior research (Crossley 2018), the SMs that included at least 150 words were aggregated to create a single file for each physician. A total of 2.7% physicians from the overall set were filtered out because their aggregated SMs were less than 150 words. A sufficient number of words are necessary to have a stable measure of text complexity and hence the decision to use 150-word threshold. Future work may indeed lead to algorithms that require fewer words; however, using a minimum number of words follows standard practice in the readability literature. To be consistent and to be able to examine concordance between the patient HL and the physician CP in the future, only those SMs written by physicians to patients were included, whose SMs were used in an earlier analysis (Crossley et al. 2020). It was observed that while using the electronic health record portal some physicians made elaborate use of "automated text" that is available to them while they draft their responses. These automatically generated texts in the SMs were retained for linguistic analyses because it proved difficult to create an automatic and generic NLP/ML model to reliably exclude these automated text segments, and because these texts represented the language used by the physicians when replying to their patients.

10.9.2 EXPERT RATING

The readability of the physicians' SMs sent to the patients were rated by two experts based on the difficulty faced by "struggling readers" in processing and comprehending the SMs for (Protection and Act 2010; Schillinger et al. 2017). Both of the raters were experts in medical discourse, had taught classes in literacy, and had experience evaluating/rating medical texts. The raters were instructed to use a 5-point Likert scale scoring rubric. The raters were required to judge "the ease with which a struggling reader could understand the message?" The 5-point Likert scale ranged from: 1-"very easy" to 5-"very difficult". This approach used was similar to Kandula and Zeng-Treitler's (2008) 7-point scale rating.

10.9.3 CREATING PHYSICIAN COMPLEXITY PROFILE: MoTeR-P

The Model of Text Readability in Physicians (MoTeR-P) was developed from the SMs written by the physicians to their patients and the expert scores. NLP tools similar to those discussed in Section 10.6 were employed to compute each SM linguistic features. The linguistic features used to develop the CP were similar to text complexity measures that had been previously validated. TAALES and SÉANCE were used to compute the features related to lexical sophistication and semantics. Syntactic complexity-related features were calculated using TAASSC (Kyle 2016, and Coh-Metrix (Graesser et al. 2004). TAACO was used for text cohesion features. Finally, Coh-Metrix was used to return the traditional readability scores (FKGL and FRE).

SMs scored by experts in the range of 1–3 were classified as "low CP" (i.e., easy to understand), while those scored 4–5 were classified as "high CP" (i.e., difficult). The linguistic features were checked and pruned for non-normality based on skewness and kurtosis values (George 2011), effect size with the dependent variable, and multicollinearity prior to the statistical analyses. The remaining 85 features were used to develop MoTeR-P for predicting text complexity of expert ratings, by using these features as predictors in training a linear discriminant analysis (LDA) model. To account for the unbalanced data (more texts scored rated as 3 i.e., neither difficult nor easy), the LDA model probability weights were adjusted such that instead of using the default 0.50 probability, any weights greater than 0.55 were classified as "easy". To control for overfitting, a threshold of one predictor per 20 instances (i.e., 24 predictors) was used to train the LDA model. For comparison, another separate LDA model was trained using the FKGL for predicting the expert ratings of text complexity.

10.9.4 EVALUATING PHYSICIAN COMPLEXITY PROFILE

The MoTeR-P used a set of 24 linguistic features for training (see Table 10.4) and achieved an accuracy of 0.749, sensitivity and specificity of 0.674 and 0.788 respectively for the test data, $X^2 = 50.977$, $p < 0.001$, and 0.455 Cohen's Kappa indicating a moderate agreement. It was observed that the SMs predicted as difficult contained words that were less familiar, more abstract, high age of acquisition, occurred in fewer texts, more frequent academic function words, fewer actions and objects words, and were more diverse (lexically). Conversely, the SMs containing more frequent tri-grams that were also present in more texts were predicted as more readable. Syntactically, SMs having fewer syntactic overlap and more dependencies across sentences were classified as harder to understand. In addition, more difficult texts contained greater number of function words, across paragraphs verb overlaps

TABLE 10.4

Descriptive statistics of the features used in MoTeR-P (Crossley et al. 2020)

Feature	Mean (SD): Difficult	Mean (SD): Easy	Feature	Mean (SD): Difficult	Mean (SD): Easy
Word range score, CW, SUBTLEXus	3.196 (0.172)	3.303 (0.123)	Average direct object dependencies	1.253 (0.368)	1.148 (0.355)
Word age of acquisition scores, AW, Kuperman	5.496 (0.39)	5.284 (0.26)	Friends and family related words	0.191 (0.081)	0.217 (0.095)
Lexical diversity (D)	89.827 (23.659)	77.259 (23.543)	Argument overlap (binary), paragraphs	0.579 (0.197)	0.524 (0.203)
FW overlap, paragraphs	8.225 (4.016)	6.379 (3.936)	Words related to joy	0.617 (0.501)	0.788 (0.68)
FW frequency, COCA academic	16831.399 (3494.824)	15496.32 (3257.931)	Argument overlap, sentences	0.219 (0.078)	0.241 (0.09)
Lexical diversity (Maas)	0.021 (0.003)	0.023 (0.004)	Trigram range, COCA fiction	0.028 (0.008)	0.026 (0.007)
Bigram association strength (delta p), COCA academic	0.045 (0.012)	0.041 (0.01)	Trigram frequency, COCA newspaper	0.429 (0.118)	0.401 (0.111)
Terms related to action	0.609 (0.133)	0.655 (0.142)	Pronoun overlap, paragraphs	1.354 (0.66)	1.199 (0.648)
Syntactic similarity score	0.104 (0.031)	0.116 (0.039)	Construction frequency, SD, COCA fiction	650503.988 (135651.74)	620837.079 (139117.53)
Word familiarity, AW, MRC	593.239 (4.894)	594.622 (3.825)	Word concreteness, AW, MRC	2.642 (0.136)	2.669 (0.126)
Fear and disgust words	0.118 (0.067)	0.097 (0.07)	Incidence of words related to objects	0.134 (0.053)	0.145 (0.061)
Verb overlap (binary), paragraphs	0.764 (0.257)	0.674 (0.311)	Argument overlap, paragraphs	0.383 (0.159)	0.413 (0.164)

AW: all words; CW: content words; FW: function words; SD: standard deviation.

and fewer semantic overlaps, when cohesion was considered. From sentiment perspective, messages containing more words expressing fear and disgust, and fewer words associated with joy, friends, and family were classified as more difficult to understand.

The MoTeR-P model's performance, when compared with the FKGL model, was found to be superior. The FKGL-LDA model for the same dataset reported an accuracy of 0.65, X^2 6.56, p = 0.01, and demonstrated weak agreement with 0.154 Cohen's Kappa. Sensitivity and specificity for the FKGL model were found to be 0.302 and 0.838 respectively, which were much lower than those of the MoTeR-P.

10.10 CHALLENGES AND SOLUTIONS

Like any other project, the ECLIPPSE project team during implementation encountered and overcame several challenges. This section describes the challenges encountered (while employing NLP/ML techniques for developing/validating the LPs and CPs already discussed in the previous sections of this chapter), and the solutions devised to address these challenges (see also Brown et al. 2021). The challenges and the solutions are broadly categorized into a) data mining-related issues, b) linguistic indices analyses-based problems, and c) interdisciplinary collaboration-related complications.

10.10.1 SMs CORPUS MINING

The first challenge was related to data mining of the SMs. The SMs for both the patient and physician in isolation and the patient-physician interactions had to be extracted. The data had to be extracted from multiple locations and mapped according to their identification numbers (IDs). For this, the patients' unique IDs referred to as the medical record numbers (MRNs) from their EHRs had to be matched to their IDs assigned by the KP patient portal. In turn, the message IDs of the KP patient portal had to be mapped to their message IDs in the EHR data to extract the SMs from their EHR notes. Subsequent to the data extraction, there were challenges related to missing or incorrect paragraph separators, sentence boundary markers, and punctuations (often referred to as structural markers). The absence of such markers influenced computation of several linguistic features across paragraphs, syntactic indices, potentially leading to imprecise computations and incorrect parsing.

The problems related to parsing were further compounded by the presence of some unstructured and ungrammatical contents in the SMs including test reports from labs, website links, automated signatures, office addresses along with their office hours, etc. Another issue that necessitated additional data security measures was the presence of patient/physician information such as their names and phone numbers in some SMs. It was not feasible to correctly de-identify all the data using automatic measures, therefore the data needed to be stored and could only be analyzed on secure servers using KPNC firewalls. While storing the data and analyzing it on the secure KPNC server did solve the issue related to the confidentiality of data and security, but this also resulted in challenges related to server accessibility, and delays in data processing.

In addition, physicians' SMs often included "smart texts" or "smart phrases". These texts/phrases refer to pre-defined automated content made available to the physicians by the online KP portal while the physicians respond to their patients. As described in section 10.9.1, these automated smart texts were not removed from the corpus because of large number of variations in the smart texts and phrases, it was infeasible to create an NLP algorithm that could generalize well and automatically identify all these varying texts/phrases with high accuracy. Another reason for retaining these automated texts was that these texts represented the language that was actually used by physicians while writing to their patients and would form an important and integral part of subsequent linguistic analyses.

Beyond the data mining problems, there were also issues on the patient side regarding who was the author of the SMs being written to the physician. It was observed sometimes that the SMs instead of being written by the patient were written by the patient proxies. The "ProxyID" algorithm was developed and used to identify hidden proxy messages (Semere et al. 2019). In addition, some SMs contained text that was in language other than English. Due to limited availability of NLP tools for non-English texts, scripts were created to identify such texts that successfully removed non-English (Spanish, in specific) texts if more than 50% of the text was in a language other than English. Due to this threshold for removing texts, some residual non-English text may have been left in the SMs.

10.10.2 Linguistic Indices Analyses

It is not always feasible to perform robust linguistic analysis if the SMs to be analyzed are short. We encountered some SMs in the corpus that were too short, leading to structural challenges in analyzing the data. In order to address this issue of content in an SM being insufficient to be linguistically analyzed correctly, a minimum word requirement was applied to the SMs. Those SMs written by patient containing fewer than 50 words were removed from analysis, and only SMs with more than 150 words written by physicians were considered for future analysis.

Selecting best features from a set of linguistic indices for training the machine learning models is always a challenge. We needed to select a set of linguistic indices for the LP and CP algorithms from a set of hundreds returned by different linguistic tools. As a result, commonly used filter methods such as multi-collinearity, zero, and nearly zero variance and non-normal distributions were used to filter out some linguistic features before the model was trained. The model was trained using the indices obtained after filtering to identify some of the topmost important indices within the trained model to further reduce the linguistic indices. Imbalanced samples in HL estimations were another matter of concern while developing the LP models. Some traditional ML algorithms are not designed to handle skewed or imbalanced data and hence do not perform well with such data. Several methods (e.g., under-sampling, oversampling, or SMOTE) were examined to account for the imbalance in the datasets and different measures (such as updating thresholds, refining expert ratings) were explored to handle such data. The computational processes and the validity for these measures are detailed in papers describing the LP (Balyan et al. 2019; Crossley et al. 2021; Schillinger et al. 2020) and CP (Crossley et al. 2020) development.

Another critical challenge faced by the team was assessing the LP and CP performance in the absence of "gold standards" for HL and linguistic complexity for the patient and physician respectively. One of the solutions included expert ratings of SMs for a small subset of the overall data. A challenge related to the expert ratings was the need to refine the scoring rubrics and reliably train the raters for assessing the HL and linguistic complexity of both the patient and the physician. The gold standard problem for the LP was overcome by applying two proxy measures: DISTANCE survey self-reported HL (Sarkar et al. 2010) and an expert-rated HL measure. As a result, two versions of the LPs were generated (LP-SR: Balyan et al. 2019; Schillinger et al. 2020; and LP-Exp: Crossley et al. 2021). For the physician CP, because of non-availability of a true gold standard, a measure was developed based on the expert ratings of the physician SM linguistic complexity (Crossley et al. 2020).

10.10.3 INTERDISCIPLINARY COLLABORATION

Experts working across several scientific disciplines and geographies encountered various challenges, such as similar terms across different domains (health services research, linguistics, and cognitive science) had different meanings, definitional differences, and lack of understanding related to tasks and methods at times resulted in confusion and inefficiency. A few more critical transdisciplinary challenges were related to methods transportability, research integrity or rigor, different interpretations of certain findings in terms of their real-world significance, and agreeing on priorities related to research and publications.

Challenges inherent to interdisciplinary collaboration were addressed by employing real-time and eventual clarification, proper documentation of ambiguous terms and tasks. Annual in-person meetings, biweekly video conferences, and regular email exchanges speed up decision making, resolve discrepancies related to terminologies, and ensured consistency and consensus-building. Providing background and context to align objectives and clarifying discipline-specific methodologies and terminologies also proved helpful. Even though more discussions were needed for some tasks, frequent delineation and re-visiting the grant aims helped mitigate tensions between different aspects of the project (theoretical vs. applied).

10.11 CONCLUSION

This chapter discussed the processes for development, evaluation, validation, and various challenges encountered and solutions devised for patients' LPs and physicians' CPs, the scientific products of the first two aims of the ECLIPPSE project. While a variety of healthcare research applications have employed NLP and ML approaches, the research discussed in this Chapter is the first of its kind to classify patients' HL and physician language complexity using the SMs exchanged between patients and physicians. Nonetheless, the studies have several limitations that should be noted.

First, in the context of patients' LP, only patients who wrote SMs were analyzed, which may have excluded patients with severe HL limitations. Second, the study was limited to SMs written in English only, excluding other patients with limited HL. Third, the LPs were modeled against HL that was self-reported by patients and rated by experts, due to non-availability of a comprehensive gold standard for HL; this may have limited the classification of HL. Fourth, this study analyzed data from a large and integrated healthcare system; hence, how well the models work in other systems and for patients other than DM2 patients needs to be assessed. Finally, adding patient demographics and clinical characteristics to the current linguistic features from SMs could increase the accuracy of our models.

Limitations of the study dealing with physicians' CP also exist. First, the current linguistic features did not include indicators of interactivity, shared meaning, and empathy, and hence may be insufficient in predicting communication-sensitive outcomes. Second, only linguistic features were used for the study; features beyond linguistics such as age, reading ability, background knowledge, and socio-cultural background were not incorporated. As such, future research needs to assess the extent to which physician CP is associated with patient comprehension, as directly measured. Third, only text content in the SMs was analyzed; other aspects, such as figures and charts were not considered. Finally, the CP work was limited to PCPs' SMs; other healthcare team members such as nurses, medical assistants, and sub-specialists were not considered.

We conclude that applying innovative NLP and ML approaches to generate a patient-physician LP and CP from their SMs is a feasible, and scalable strategy for identifying patients with limited HL, and to identify those physicians who write complex messages to their patients. This work can provide a tool that has potential to reduce disparities related to HL. Additionally, patients classified as limited HL can be provided feedback for promoting adherence (Sudore et al. 2009). It may prove useful to identify limited HL patients which may be further used to alert clinicians about their patients' potential difficulties in written and verbal instructions comprehension. Finally, identifying complex messages on the part of physicians, and providing them with feedback could enable these physicians to tailor their messages, making them more readable and easier to comprehend and act upon.

From our perspective, the NLP methods are an important contribution of this research. Indeed, the use of NLP combined with machine learning was the primary purpose of this project. The Literacy profiles derived using linguistic and semantic indices based on patients' written secure messages is an entirely novel contribution. Likewise, identifying linguistic complexity profiles based on the secure messages composed by the physicians to their patients is equally novel and a major contribution to the literature. The validity of these algorithms was assessed based on the health outcome patterns in the health domains. The work carried out in this project is innovative from an NLP application perspective. From the perspective of NLP, this is the first attempt (that we know of) to use NLP to develop literacy profiles of patients' health literacy or language complexity profiles based on physician written communication. Equally novel is the alignment of the LP and CP profiles with demographic and patient outcomes. The contribution of this chapter is that it is the only documented work that provides an overview of the project, including both profiles and information on the work involving natural language processing.

10.12 FUTURE DIRECTIONS

Effective communication between a patient and a physician is innate to clinical practice, and important for delivering high-quality healthcare. In patient-physician communication, the *manner* in which information is communicated by a physician to a patient is as *important* as the information being communicated (Travaline, Ruchinskas, and D'Alonzo 2005). A large number of patient-physician relationships breakdown due to patients' dissatisfaction because many physicians overestimate their communication abilities (Ha and Longnecker 2010). While linguistic complexity matching ("concordance") is believed to promote shared understanding, no study has tested this hypothesis. We apply HL and linguistic complexity measures for patients' and physicians' respectively generated via computational linguistics to determine whether linguistic matching has clinical benefits, and for whom. We plan to classify physician-patient dyads as concordant or discordant, and explore whether concordance is associated with the patients' reports of physician communication (CAHPS), specifically using the items that ask patients about the extent to which their doctor explains things in ways that they can understand. We will further characterize each physicians' communication "style" (the predominant strategy they employ across patients), specifically exploring whether "universal precautions" (using simpler language with all patients) or "universal tailoring" (using language that matches their patients' HL) is associated with better understanding.

One of the final aims of the ECLIPPSE project is the development and evaluaton of an online, automated feedback prototype embedded in the patient portal. We plan to test the effects of this automated feedback tool with respect to its ability to reduce the physicians' SMs complexity in order to meet the low HL patients' needs. One of our objectives is to develop an interface for physicians where they can respond to messages received from patients. For SMs that are calculated to be overly complex relative to the patients' HL, physician participants will receive different versions of linguistically motivated feedback. This SM application, if successful and found to be acceptable to physicians, could have substantial benefits for their patients.

ACKNOWLEDGMENTS

This work has been supported by grants NLM R01 LM012355 from the National Institutes of Health, NIDDK Centers for Diabetes Translational Research (P30 DK092924), R01 DK065664, NICHD R01 HD46113, Institute of Education Sciences, U.S. Department of Education, through grant R305A180261 and Office of Naval Research grant (N00014–17-1–2300).

NOTES

1 linguisticanalysistools.org
2 https://www.i2b2.org/NLP/DataSets/Main.php

REFERENCES

Baayen, R. H., R. Piepenbrock, and L. Gulikers 1996. The CELEX lexical database (cd-rom).

Bailey, S. C., A. G.Brega, T. M.Crutchfield, T. Elasy, H. Herr, K. Kaphingst, ... and D. Schillinger 2014. Update on health literacy and diabetes. *The Diabetes Educator* 40, no. 5: 581–604.

Badarudeen, S., and S. Sabharwal 2010. Assessing readability of patient education materials: Current role in orthopaedics. *Clinical Orthopaedics and Related Research®* 468, no. 10: 2572–2580.

Baker, D. W., D. A. DeWalt, D. Schillinger, V. Hawk, B. Ruo, K. Bibbins-Domingo, ... and M. Pignone 2011. "Teach to goal": Theory and design principles of an intervention to improve heart failure self-management skills of patients with low health literacy. *Journal of Health Communication* 16, no. sup3: 73–88.

Balyan, R., S. A. Crossley, W. Brown III, A. J. Karter, D. S. McNamara, J. Y. Liu, ... and D. Schillinger 2019. Using natural language processing and machine learning to classify health literacy from secure messages: The ECLIPPSE study. *PloS One* 14, no. 2: e0212488.

Balyan, R., K. S. McCarthy, and D. S. McNamara 2017. Combining machine learning and natural language processing to assess literary text comprehension. In A. Hershkovitz and L. Paquette (eds.). In Proceedings of the 10th International Conference on Educational Data Mining (EDM), Wuhan, China: International Educational Data Mining Society.

Balyan, R., K. S. McCarthy, and D. S. McNamara 2018. Comparing machine learning classification approaches for predicting expository text difficulty. In Proceedings of the 31st Annual Florida Artificial Intelligence Research Society International Conference (FLAIRS). AAAI Press, Florida.

Balyan, R., K. S. McCarthy, and D. S.McNamara 2020. Applying natural language processing and hierarchical machine learning approaches to text difficulty classification. *International Journal of Artificial Intelligence in Education* 30, no. 3: 337–370.

Barton, J. L., L. Trupin, D. Schillinger, G. Evans-Young, J. Imboden, V. M. Montori, and E. Yelin 2016. Use of low-literacy decision aid to enhance knowledge and reduce decisional conflict among a diverse population of adults with rheumatoid arthritis: Results of a pilot study. *Arthritis Care & Research* 68, no. 7: 889–898.

Berland, G. K., M. N. Elliott, L. S. Morales, J. I. Algazy, R. L. Kravitz, M. S. Broder, ... and E. A. McGlynn 2001. Health information on the internet: Accessibility, quality, and readability in English and Spanish. *JAMA* 285, no. 20: 2612–2621.

BNC Consortium. 2007. *British National Corpus.* University of Oxford, Oxford Text Archive Core Collection, UK.

Boulos, M. N. K. 2005. British internet-derived patient information on diabetes mellitus: Is it readable? *Diabetes Technology & Therapeutics* 7, no. 3: 528–535.

Brach, C., B. P. Dreyer, and D. Schillinger 2014. Physicians' roles in creating health literate organizations: A call to action. *Journal of General Internal Medicine* 29, no. 2: 273–275.

Brach, C., D. Keller, L. M. Hernandez, C. Baur, R. Parker, B. Dreyer, ... and D. Schillinger 2012. Ten attributes of health literate health care organizations. *NAM Perspectives, Institute of Medicine of the National Academes.*

Breiman, L. 1996. Bagging predictors. *Machine Learning* 24, no. 2: 123–140.

Brown III, W., R. Balyan, A. J. Karter, S. Crossley, W. Semere, N. D. Duran, ... and D. Schillinger 2021. Challenges and solutions to employing natural language processing and machine learning to measure patients' health literacy and physician writing complexity: The ECLIPPSE study. *Journal of Biomedical Informatics* 113: 103658.

Bruce, B., and A. Rubin 1988. Readability formulas: Matching tool and task. Lawrence Erlbaum Associates, Inc.

Bruce, B., A. Rubin, and K. Starr 1981. Why readability formulas fail. *IEEE Transactions on Professional Communication*, no.1: 50–52.

Castro, C. M., C. Wilson, F. Wang, and D. Schillinger 2007. Babel babble: Physicians' use of unclarified medical jargon with patients. *American Journal Of Health Behavior* 31, no. 1: S85–S95.

Chall, J. S., and E. Dale 1995. *Readability Revisited: The New Dale-Chall Readability Formula*. Brookline Books.

Chew, L. D., J. M. Griffin, M. R. Partin, S. Noorbaloochi, J. P. Grill, A. Snyder, … and M. VanRyn 2008. Validation of screening questions for limited health literacy in a large VA outpatient population. *Journal of General Internal Medicine* 23, no. 5: 561–566.

Coltheart, M. 1981. The MRC psycholinguistic database. *The Quarterly Journal of Experimental Psychology Section A* 33, no. 4: 497–505.

Crossley, S. A. 2018. How many words needed? Using natural language processing tools in educational data mining. In Proceedings of the 10th International Conference on Educational Data Mining (EDM). pp. 630–633.

Crossley, S. A., R. Balyan, J. Liu, A. J. Karter, D. McNamara, and D. Schillinger 2021. Developing and testing automatic models of patient communicative health literacy using linguistic features: Findings from the ECLIPPSE study. *Health Communication* 36, no. 8: 1018–1028.

Crossley, S. A., R. Balyan, J. Liu, A. J. Karter, D. McNamara, and D. Schillinger 2020. Predicting the readability of physicians' secure messages to improve health communication using novel linguistic features: Findings from the ECLIPPSE study. *Journal of Communication in Healthcare* 13, no. 4: 344–356.

Crossley, S. A., J. Greenfield, and D. S. McNamara 2008. Assessing text readability using cognitively based indices. *Tesol Quarterly* 42, no. 3: 475–493.

Crossley, S., and V. Kostyuk 2017. Letting the genie out of the lamp: Using natural language processing tools to predict math performance. In International Conference on Language, Data and Knowledge. pp. 330–342. Springer, Cham.

Crossley, S. A., K. Kyle, and D. S. McNamara 2015. To aggregate or not? Linguistic features in automatic essay scoring and feedback systems. *Grantee Submission* 8, no. 1. Retrieved from https://escholarship.org/uc/item/1f21q8ck

Crossley, S. A., K. Kyle, and D. S. McNamara 2016. The tool for the automatic analysis of text cohesion (TAACO): Automatic assessment of local, global, and text cohesion. *Behavior Research Methods* 48, no. 4: 1227–1237.

Crossley, S. A., K. Kyle, and D. S. McNamara 2017. Sentiment analysis and social cognition engine (SEANCE): An automatic tool for sentiment, social cognition, and social-order analysis. *Behavior Research Methods* 49, no. 3: 803–821.

Crossley, S., L. Paquette, M. Dascalu, D. S. McNamara, and R. S. Baker 2016. Combining click-stream data with NLP tools to better understand MOOC completion. In Proceedings of the Sixth International Conference on Learning Analytics & Knowledge. pp. 6–14.

Crossley, S., R. Roscoe, and D. McNamara 2013. Using automatic scoring models to detect changes in student writing in an intelligent tutoring system. In the Twenty-Sixth International FLAIRS Conference.

Crossley, S. A., S. Skalicky, M. Dascalu, D. S. McNamara, and K. Kyle 2017. Predicting text comprehension, processing, and familiarity in adult readers: New approaches to readability formulas. *Discourse Processes* 54, no. 5-6: 340–359.

Davison, A., and R. N. Kantor 1982. On the failure of readability formulas to define readable texts: A case study from adaptations. *Reading Research Quarterly*: 187–209.

De Clercq, O., V. Hoste, B. Desmet, P. Van Oosten, M. De Cock, and L. Macken 2014. Using the crowd for readability prediction. Natural *Language Engineering* 20, no. 3: 293–325.

De Marneffe, M. C., B. MacCartney, and C. D. Manning 2006. Generating typed dependency parses from phrase structure parses. *International Conference on Language Resources and Evaluation* 6: 449–454.

DeWalt, D. A., D. W. Baker, D. Schillinger, V. Hawk, B. Ruo, K. Bibbins-Domingo, ... and M. Pignone 2011. A multisite randomized trial of a single-versus multi-session literacy sensitive self-care intervention for patients with heart failure. *Journal of General Internal Medicine* 26: S57–S58. 233 Spring st, New York, 10013 USA: Springer.

DeWalt, D. A., K. A. Broucksou, V. Hawk, D. W. Baker, D. Schillinger, B. Ruo, ... & M. Pignone 2009. Comparison of a one-time educational intervention to a teach-to-goal educational intervention for self-management of heart failure: Design of a randomized controlled trial. *BMC Health Services Research* 9, no. 1: 1–14.

Dumais, S., J. Platt, D. Heckerman, and M. Sahami 1998. Inductive learning algorithms and representations for text categorization. In Proceedings of the Seventh International Conference on Information and Knowledge Management. pp. 148–155.

Fisher, R. A. 1936. The use of multiple measurements in taxonomic problems. *Annals of Eugenics* 7, no. 2: 179–188.

Flesch, R. 1948. A new readability yardstick. *Journal of Applied Psychology* 32, no. 3: 221.

Friedman, D. B., and L. Hoffman-Goetz 2006. A systematic review of readability and comprehension instruments used for print and web-based cancer information. *Health Education & Behavior* 33, no. 3: 352–373.

Fry, E. 1968. A readability formula that saves time. *Journal of Reading* 11, no. 7: 513–578.

Gemoets, D., G. Rosemblat, T. Tse, and R. A. Logan 2004. Assessing readability of consumer health information: An exploratory study. *Medinfo*. pp. 869–873.

George, D. 2011. SPSS for windows step by step: A simple study guide and reference. *17.0 Update, 10/e*. Pearson Education India.

Grabeel, K. L., J. Russomanno, S. Oelschlegel, E. Tester, and R. E. Heidel 2018. Computerized versus hand-scored health literacy tools: A comparison of Simple Measure of Gobbledygook (SMOG) and Flesch-Kincaid in printed patient education materials. *Journal of the Medical Library Association: JMLA* 106, no. 1: 38.

Graesser, A. C., D. S. McNamara, and J. M. Kulikowich 2011. Coh-Metrix: Providing multilevel analyses of text characteristics. *Educational Researcher* 40, no. 5: 223–234.

Graesser, A. C., D. S. McNamara, M. M. Louwerse, and Z. Cai 2004. Coh-Metrix: Analysis of text on cohesion and language. *Behavior Research Methods, Instruments, & Computers* 36, no. 2: 193–202.

Gunning, R. 1952. *Technique of Clear Writing*. McGraw Hill.

Ha, J. F., and N. Longnecker 2010. Doctor-patient communication: A review. *Ochsner Journal* 10, no. 1: 38–43.

Harris, L. T., T. D. Koepsell, S. J. Haneuse, D. P. Martin, and J. D. Ralston 2013. Glycemic control associated with secure patient-provider messaging within a shared electronic medical record: A longitudinal analysis. *Diabetes Care* 36, no. 9: 2726–2733.

Hastie, T., R. Tibshirani, and J. Friedman 2009. *The Elements of Statistical Learning: Data Mining, Inference, and Prediction*. Springer Science & Business Media.

Hill-Briggs, F., K. P. Schumann, and O. Dike 2012. 5-step methodology for evaluation and adaptation of print patient health information to meet the < 5th grade readability criterion. *Medical Care* 50, no. 4: 294.

Hudson, S., R. V. Rikard, I. Staiculescu, and K. Edison 2018. Improving health and the bottom line: The case for health literacy. In Building the Case for Health Literacy: Proceedings of a Workshop. National Academies Press (US).

Institute of Medicine. 2004. *Health lIteracy: A Prescription to End Confusion*. Washington, DC: The National Academies Press.

James, G., D. Witten, T. Hastie, and R. Tibshirani 2013. *An Introduction to Statistical Learning*, Vol. 112, p. 18. New York: Springer.

Jindal, P., and J. C. MacDermid 2017. Assessing reading levels of health information: Uses and limitations of flesch formula. *Education for Health* 30, no. 1: 84.

Joachims, T. 1998. Text categorization with support vector machines: Learning with many relevant features. In *European Conference on Machine Learning*. pp. 137–142. Berlin, Heidelberg: Springer.

Johnson, S. B. 1999. A semantic lexicon for medical language processing. *Journal of the American Medical Informatics Association* 6, no. 3: 205–218.

Kandula, S., and Q. Zeng-Treitler 2008. Creating a gold standard for the readability measurement of health texts. In *AMIA Annual Symposium Proceedings*, Vol. 2008, p. 353. American Medical Informatics Association.

Karter, A. J., M. M. Parker, O. K. Duru, D. Schillinger, N. E. Adler, H. H. Moffet, ... and J. A. Schmittdiel 2015. Impact of a pharmacy benefit change on new use of mail order pharmacy among diabetes patients: The Diabetes Study of Northern California (DISTANCE). *Health Services Research* 50, no. 2: 537–559.

Karter, A. J., M. M. Parker, H. H. Moffet, A. T. Ahmed, J. A. Schmittdiel, and J. V. Selby 2009. New prescription medication gaps: A comprehensive measure of adherence to new prescriptions. *Health Services Research* 44, no. 5p1: 1640–1661.

Kim, H., S. Goryachev, G. Rosemblat, A. Browne, A. Keselman, and Q. Zeng–Treitler 2007. Beyond surface characteristics: A new health text-specific readability measurement. In *AMIA Annual Symposium Proceedings*. pp. 418–422. Chicago, IL: American Medical Informatics Association.

Kincaid, J. P., R. P. Fishburne, Jr, R. L. Rogers, and B. S. Chissom 1975. Derivation of new readability formulas (automated readability index, fog count and flesch reading ease formula) for navy enlisted personnel. Naval Technical Training Command Millington TN Research Branch.

Kindig, D. A., A. M. Panzer, and L. Nielsen-Bohlman (Eds.). 2004. Health literacy: A prescription to end confusion. National Academies Press.

Kirsch, I. S., A. Jungeblut, L. Jenkins, and A. Kolstad 2002. *Adult Literacy in America: A First Look at the Findings of the National Adult Literacy Survey (NCES 1993–275)*. Washington, DC: U.S. Department of Education.

Krogh, A., and J. Vedelsby 1995. Neural network ensembles, cross validation, and active learning. In *Advances in Neural Information Processing Systems*. pp. 231–238.

Kugar, M. A., A. C. Cohen, W. Wooden, S. S. Tholpady, and M. W. Chu 2017. The readability of psychosocial wellness patient resources: Improving surgical outcomes. *Journal of Surgical Research* 218: 43–48.

Kusec, S., O. Brborovic, and D. Schillinger 2003. Diabetes websites accredited by the Health on the Net Foundation Code of Conduct: Readable or not? *Studies in Health Technology and Informatics* 95: 655–660.

Kyle, K. 2016. Measuring syntactic development in L2 writing: Fine grained indices of syntactic complexity and usage-based indices of syntactic sophistication. Dissertation, Georgia State University.

Kyle, K., and S. A. Crossley 2015. Automatically assessing lexical sophistication: Indices, tools, findings, and application. *Tesol Quarterly* 49, no. 4: 757–786.

Kyle, K., S. Crossley, and C. Berger 2018. The tool for the automatic analysis of lexical sophistication (TAALES): Version 2.0. *Behavior Research Methods* 50, no. 3: 1030–1046.

Lu, X. 2010. Automatic analysis of syntactic complexity in second language writing. *International Journal of Corpus Linguistics* 15, no. 4: 474–496.

Lyles, C. R., U. Sarkar, D. Schillinger, J. D. Ralston, J. Y. Allen, R. Nguyen, and A. J. Karter 2016. Refilling medications through an online patient portal: Consistent improvements in adherence across racial/ethnic groups. *Journal of the American Medical Informatics Association* 23, no. e1: e28–e33.

Malvern, D., B. Richards, N. Chipere, and P. Durán 2004. *Lexical Diversity and Language Development*. New York: Palgrave Macmillan.

McAndie, E., A. Gilchrist, and B. Ahamat 2016. Readability of clinical letters sent from a young people's department. *Child and Adolescent Mental Health* 21, no. 3: 169–174.

McCallum, A., and K. Nigam 1998. A comparison of event models for naive bayes text classification. *AAAI-98 Workshop on Learning for Text Categorization* 752, no. 1: 41–48.

McCarthy, P. M. 2005. *An aSsessment of the Range and Usefulness of Lexical Diversity Measures and the Potential of the Measure of Textual, Lexical Diversity (MTLD)* Doctoral dissertation, Tennessee: The University of Memphis.

McLaughlin, G. H. 1969. SMOG grading-a new readability formula. *Journal of Reading* 12, no. 8: 639–646.

McNamara, D. S., S. A. Crossley, and P. M. Mccarthy 2010. Linguistic features of writing quality. *Written Communication* 27, no. 1: 57–86.

McNamara, D. S., S. A. Crossley, and R. Roscoe 2013. Natural language processing in an intelligent writing strategy tutoring system. *Behavior Research Methods* 45, no. 2: 499–515.

McNamara, D. S., S. A. Crossley, R. D. Roscoe, L. K. Allen, and J. Dai 2015. A hierarchical classification approach to automated essay scoring. *Assessing Writing* 23: 35–59.

McNamara, D. S., A. C. Graesser, P. M. McCarthy, and Z. Cai 2014. *Automated Evaluation of Text and Discourse with Coh-Metrix*. Cambridge University Press.

McNamara, D. S., R. Roscoe, L. K. Allen, R. Balyan, and K. S. McCarthy 2019. Literacy: From the perspective of text and discourse theory. *Journal of Language and Education* 5, no. 3: 56–69.

Meade, C. D., J. C. Byrd, and M. Lee 1989. Improving patient comprehension of literature on smoking. *American Journal of Public Health* 79, no. 10: 1411–1412.

Miller, G. A. 1995. WordNet: A lexical database for English. *Communications of the ACM* 38, no. 11: 39–41.

Mitchell, T. M. 1997. *Machine Learning*. New York: McGraw-Hill.

Moffet, H. H., N. Adler, D. Schillinger, A. T. Ahmed, B. Laraia, J. V. Selby, … and A. J. Karter 2009. Cohort profile: The Diabetes Study of Northern California (DISTANCE)—Objectives and design of a survey follow-up study of social health disparities in a managed care population. *International Journal of Epidemiology* 38, no. 1: 38–47.

Munsour, E. E., A. Awaisu, M. A. A. Hassali, S. Darwish, and E. Abdoun 2017. Readability and comprehensibility of patient information leaflets for antidiabetic medications in Qatar. *Journal of Pharmacy Technology* 33, no. 4: 128–136.

Oliffe, M., E. Thompson, J. Johnston, D. Freeman, H. Bagga, and P. K. Wong 2019. Assessing the readability and patient comprehension of rheumatology medicine information sheets: A cross-sectional health literacy study. *BMJ Open* 9, no. 2: e024582.

Paasche-Orlow, M. K., H. A. Taylor, and F. L. Brancati 2003. Readability standards for informed-consent forms as compared with actual readability. *New England Journal of Medicine* 348, no. 8: 721–726.

Piñero-López, M. Á., P. Modamio, C. F. Lastra, and E. L. Mariño 2016. Readability analysis of the package leaflets for biological medicines available on the internet between 2007 and 2013: An analytical longitudinal study. *Journal of Medical Internet Research* 18, no. 5: e100.

Pitler, E., and A. Nenkova 2008. Revisiting readability: A unified framework for predicting text quality. In Proceedings of the 2008 conference on empirical methods in natural language processing. pp. 186–195.

Protection, P., and A. C. Act 2010. Patient protection and affordable care act. *Public Law* 111, no. 48: 759–762.

Ratanawongsa, N., A. J. Karter, M. M. Parker, C. R. Lyles, M. Heisler, H. H. Moffet, ... and D. Schillinger 2013. Communication and medication refill adherence: The diabetes study of Northern California. *JAMA Internal Medicine* 173, no. 3: 210–218.

Reed, M., J. Huang, R. Brand, I. Graetz, R. Neugebauer, B. Fireman, ... and J. Hsu 2013. Implementation of an outpatient electronic health record and emergency department visits, hospitalizations, and office visits among patients with diabetes. *Jama* 310, no. 10: 1060–1065.

Reed, M., J. Huang, I. Graetz, R. Brand, J. Hsu, B. Fireman, and M. Jaffe 2012. Outpatient electronic health records and the clinical care and outcomes of patients with diabetes mellitus. *Annals of Internal Medicine* 157, no. 7: 482–489.

Rojas, R. 2013. *Neural Networks: A Systematic Introduction*. Springer Science & Business Media.

Rubin, A. 1985. How useful are readability formulas. *Reading Education: Foundations for a Literate America*: 61–77.

Sarkar, U., A. J. Karter, J. Y. Liu, H. H. Moffet, N. E. Adler, and D. Schillinger 2010. Hypoglycemia is more common among type 2 diabetes patients with limited health literacy: The diabetes study of Northern California (DISTANCE). *Journal of General Internal Medicine* 25, no. 9: 962–968.

Sarkar, U., C. R. Lyles, M. M. Parker, J. Allen, R. Nguyen, H. H. Moffet, ... and A. J. Karter 2014. Use of the refill function through an online patient portal is associated with improved adherence to statins in an integrated health system. *Medical Care* 52, no. 3: 194.

Sarkar, U., J. D. Piette, R. Gonzales, D. Lessler, L. D. Chew, B. Reilly, ... and D. Schillinger 2008. Preferences for self-management support: Findings from a survey of diabetes patients in safety-net health systems. *Patient Education and Counseling* 70, no. 1: 102–110.

Sarkar, U., D. Schillinger, A. López, and R. Sudore 2011. Validation of self-reported health literacy questions among diverse English and Spanish-speaking populations. *Journal of General Internal Medicine* 26, no. 3: 265–271.

Schapire, R. E., and Y. Singer 2000. BoosTexter: A boosting-based system for text categorization. *Machine Learning* 39, no. 2-3: 135–168.

Schillinger, D. 2007. Literacy and health communication: Reversing the 'inverse care law'. *The American Journal of Bioethics* 7, no. 11: 15–18.

Schillinger, D., R. Balyan, S. Crossley, D. McNamara, and A. Karter 2021. Validity of a computational linguistics-derived automated health literacy measure across race/ethnicity: Findings from the ECLIPPSE project. *Journal of Health Care for the Poor and Underserved* 32, no. 2: 347–365.

Schillinger, D., R. Balyan, S. A. Crossley, D. S. McNamara, J. Y. Liu, and A. J. Karter 2020. Employing computational linguistics techniques to identify limited patient health literacy: Findings from the ECLIPPSE study. *Health Services Research*. 56, no. 1: 132–144.

Schillinger, D., A. Bindman, F. Wang, A. Stewart, and J. Piette 2004. Functional health literacy and the quality of physician–patient communication among diabetes patients. *Patient Education and Counseling* 52, no. 3: 315–323.

Schillinger, D., K. Grumbach, J. Piette, F. Wang, D. Osmond, C. Daher, ... and A. B. Bindman 2002. Association of health literacy with diabetes outcomes. *Jama* 288, no. 4: 475–482.

Schillinger, D., H. Hammer, F. Wang, J. Palacios, I. McLean, A. Tang, ... and M. Handley 2008. Seeing in 3-D: Examining the reach of diabetes self-management support strategies in a public health care system. *Health Education & Behavior* 35, no. 5: 664–682.

Schillinger, D., M. Handley, F. Wang, and H. Hammer 2009. Effects of self-management support on structure, process, and outcomes among vulnerable patients with diabetes: A three-arm practical clinical trial. *Diabetes Care* 32, no. 4: 559–566.

Schillinger, D., D. McNamara, S. Crossley, C. Lyles, H. H. Moffet, U. Sarkar, ... and A. J. Karter 2017. The next frontier in communication and the ECLIPPSE study: Bridging the linguistic divide in secure messaging. *Journal of Diabetes Research* 2017.

Schillinger, D., J. Piette, K. Grumbach, F. Wang, C. Wilson, C. Daher, ... and A. B. Bindman 2003. Closing the loop: Physician communication with diabetic patients who have low health literacy. *Archives of Internal Medicine* 163, no. 1: 83–90.

Schumaier, A. P., R. Kakazu, C. E. Minoughan, and B. M. Grawe 2018. Readability assessment of American shoulder and elbow surgeons patient brochures with suggestions for improvement. *JSES Open Access* 2, no. 2: 150–154.

Seligman, H. K., F. F. Wang, J. L. Palacios, C. C. Wilson, C. Daher, J. D. Piette, and D. Schillinger 2005. Physician notification of their diabetes patients' limited health literacy: A randomized, controlled trial. *Journal of General Internal Medicine* 20, no. 11: 1001–1007.

Semere, W., S. Crossley, A. J. Karter, C. R. Lyles, W. Brown, M. Reed, ... and D. Schillinger 2019. Secure messaging with physicians by proxies for patients with diabetes: Findings from the ECLIPPSE Study. *Journal of General Internal Medicine* 34, no. 11: 2490–2496.

Smith, F. 2012. *Understanding Reading: A Psycholinguistic Analysis of Reading and Learning to Read.* Routledge.

Smith, S. G., R. O'Conor, L. M.Curtis, K. Waite, I. J. Deary, M. Paasche-Orlow, and M. S. Wolf 2015. Low health literacy predicts decline in physical function among older adults: Findings from the LitCog cohort study. *J Epidemiol Community Health* 69, no. 5: 474–480.

Smola, A. J., and B. Schölkopf 1998. *Learning with Kernels*, Vol. 4. GMD-Forschungszentrum Informationstechnik.

Steiner, J. F., T. D. Koepsell, S. D. Fihn, and T. S. Inui 1988. A general method of compliance assessment using centralized pharmacy records: Description and validation. *Medical Care*: 814–823.

Steiner, J. F., and A. V. Prochazka 1997. The assessment of refill compliance using pharmacy records: Methods, validity, and applications. *Journal of Clinical Epidemiology* 50, no. 1: 105–116.

Sudore, R. L., C. S. Landefeld, E. J. Perez-Stable, K. Bibbins-Domingo, B. A. Williams, and D. Schillinger 2009. Unraveling the relationship between literacy, language proficiency, and patient–physician communication. *Patient Education and Counseling* 75, no. 3: 398–402.

Sudore, R. L., K. Yaffe, S. Satterfield, T. B. Harris, K. M. Mehta, E. M. Simonsick, ... and D. Schillinger 2006. Limited literacy and mortality in the elderly. *Journal of General Internal Medicine* 21, no. 8: 806–812.

Thompson, P., R. T. Batista-Navarro, G. Kontonatsios, J. Carter, E. Toon, J. McNaught, ... and S. Ananiadou 2016. Text mining the history of medicine. *PloS One* 11, no. 1: e0144717.

Travaline, J. M., R. Ruchinskas, and G. E. D'Alonzo Jr 2005. Patient-physician communication: Why and how. *Journal of the American Osteopathic Association* 105, no. 1: 13.

Uzuner, Ö. 2009. Recognizing obesity and comorbidities in sparse data. *Journal of the American Medical Informatics Association* 16, no. 4: 561–570.

Uzuner, Ö., I. Goldstein, Y. Luo, and I. Kohane 2008. Identifying patient smoking status from medical discharge records. *Journal of the American Medical Informatics Association* 15, no. 1: 14–24.

Uzuner, Ö., Y. Luo, and P. Szolovits 2007. Evaluating the state-of-the-art in automatic de-identification. *Journal of the American Medical Informatics Association* 14, no. 5: 550–563.

Uzuner, Ö., I. Solti, and E. Cadag 2010. Extracting medication information from clinical text. *Journal of the American Medical Informatics Association* 17, no. 5: 514–518.

Walsh, T. M., and T. A. Volsko 2008. Readability assessment of internet-based consumer health information. *Respiratory Care* 53, no. 10: 1310–1315.

Wang, L. W., M. J. Miller, M. R. Schmitt, and F. K. Wen 2013. Assessing readability formula differences with written health information materials: application, results, and recommendations. *Research in Social and Administrative Pharmacy* 9, no. 5: 503–516.

Wilson, M. 2009. Readability and patient education materials used for low-income populations. *Clinical Nurse Specialist* 23, no. 1: 33–40.

Wolpert, D. H. 1992. Stacked generalization. *Neural Networks* 5, no. 2: 241–259.

Wu, D. T., D. A. Hanauer, Q. Mei, P. M. Clark, L. C. An, J. Lei, … and K. Zheng 2013. Applying multiple methods to assess the readability of a large corpus of medical documents. *Studies in Health Technology and Informatics* 192: 647–651.

Zeng–Treitler, Q., S. Kandula, H. Kim, and B. Hill 2012. A method to estimate readability of health content. In Proceedings of HI-KDD 2012: ACM SICKDD Workshop on Health Informatics (HI-KDD 2012), Beijing, China.

Zhang, G. P. 2000. Neural networks for classification: A survey. *IEEE Transactions on Systems, Man, and Cybernetics, Part C (Applications and Reviews)* 30, no. 4: 451–462.

Zheng, J., and H. Yu 2017. Readability formulas and user perceptions of electronic health records difficulty: A corpus study. *Journal of Medical Internet Research* 19, no. 3: e59.

Zheng, J., and H. Yu 2018. Assessing the readability of medical documents: A ranking approach. *JMIR Medical Informatics* 6, no. 1: e17.

Zhou, Y. Y., M. H. Kanter, J. J. Wang, and T. Garrido 2010. Improved quality at Kaiser permanente through email between physicians and patients. *Health Affairs* 29, no. 7: 1370–1375.

Index

Page references followed by "n" indicate notes.